STUDENT UNIT GUIDE

NEW EDITION

WJEC AS Geography Unit G1
Changing Physical Environments

Viv Pointon

Series editor: David Burtenshaw

PHILIP ALLAN

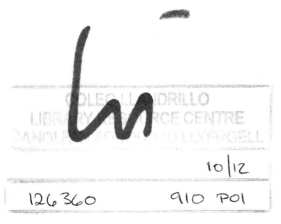

Philip Allan Updates, an imprint of Hodder Education, an Hachette UK company, Market Place, Deddington, Oxfordshire OX15 0SE

Orders
Bookpoint Ltd, 130 Milton Park, Abingdon, Oxfordshire OX14 4SB
tel: 01235 827827
fax: 01235 400401
e-mail: education@bookpoint.co.uk
Lines are open 9.00 a.m.–5.00 p.m., Monday to Saturday, with a 24-hour message answering service. You can also order through the Philip Allan Updates website: www.philipallan.co.uk

© Viv Pointon 2012

ISBN 978-1-4441-6197-7

First printed 2012
Impression number 5 4 3 2 1
Year 2016 2015 2014 2013 2012

Printed in Dubai

Hachette UK's policy is to use papers that are natural, renewable and recyclable products and made from wood grown in sustainable forests. The logging and manufacturing processes are expected to conform to the environmental regulations of the country of origin.

This material has been endorsed by WJEC and offers high quality support for the delivery of WJEC qualifications. While this material has been through a WJEC quality assurance process, all responsibility for the content remains with the publisher.

P01997

WJEC examination questions used with permission of WJEC.

Contents

Getting the most from this book

Questions & Answers

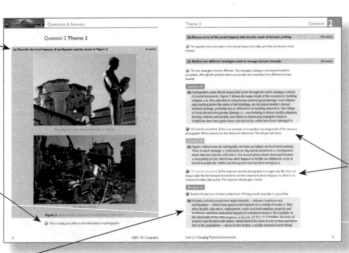

About this book

The purpose of this guide is to help you understand what is required to do well in Unit G1: Changing Physical Environments. The specification is available on the WJEC website: **www.wjec.co.uk**. This paper may be taken in Welsh. The comments all apply no matter what your language medium.

This guide is divided into two sections. The Content Guidance section gives you an introduction to the two themes that you have to study. It contains several useful diagrams and maps, which you may be able to use in the examination. The Questions & Answers section provides guidance on how to approach the unit test and includes examples of the types of question that you will see in the examination.

Content Guidance

Theme 1: Investigating climate change

1.1 What are the world's major climates and how do they relate to biomes?

The relationship between weather and climate

Weather is the atmospheric conditions that we experience from day to day — the sunshine, rain, hail, snow, temperature, wind and fog — whereas **climate** is the average weather conditions in an area over several decades. Table 1 shows climate data for Cardiff.

Table 1 Climate data for Cardiff, 1971–2000

	Jan	Feb	Mar	Apr	May	Jun	Jul	Aug	Sep	Oct	Nov	Dec	*Mean*
Mean high temperature (°C)	7	7	10	13	16	19	20	21	18	14	10	8	*13.58*
Mean low temperature (°C)	2	2	3	5	8	11	12	13	11	8	5	3	*6.92*
Mean precipitation (cm)	10.8	7.2	6.3	6.5	7.6	6.3	8.9	9.7	9.9	10.9	11.6	10.8	*8.88*

An overview of the global patterns of climate

Climate data are collated from many weather stations around the world to build up a map of global climate zones. The nature of a region's climate is determined by the temperature and precipitation it receives.

- **Temperature** depends on the amount of solar energy received, which varies spatially and seasonally. It also depends on altitude: temperature decreases with height by roughly 1°C for every 100 m above sea level.

- Mean **precipitation** is governed by global atmospheric circulation patterns that create broad bands of high rainfall at the equator and in the temperate regions, and low rainfall at the tropics and the poles.

These broad patterns (Figure 1) are modified by the Earth's rotation, which skews the patterns westwards, and by the location of the continents relative to the oceans.

Examiner tip
Don't confuse the terms 'weather' and climate'. They have different meanings, so make sure you use the correct term.

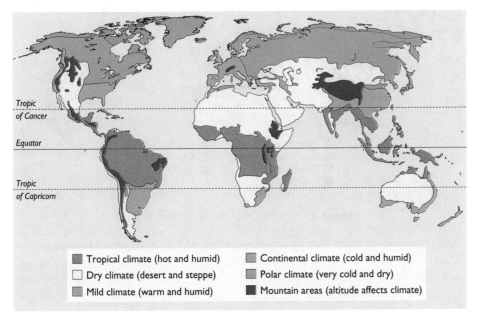

Figure 1 The world distribution of climate types

Coastal locations receive more rainfall as moisture-laden winds blow onshore. However, the proximity of the sea keeps temperatures warmer in winter and cooler in summer. Conversely, continental interiors experience much hotter summers and colder winters. This is because the specific heat capacity of the oceans is much higher than that of the land — land warms up and cools down faster. This phenomenon is called **continentality**.

An overview of biomes and their relationship with climate

The climate of a region will determine the vegetation that is able to grow there. The plant communities that survive naturally in a region form large-scale ecosystems called **biomes**. A biome is a zonal ecosystem (Figure 2). Its location is determined by climate, notably average temperature and annual precipitation. For example, if it is hot (around 27°C all year round) and wet (at least 2,000 mm of rain per year), tropical rainforest is the most likely outcome. This is described as the **climatic climax community** because it is determined by the climate. If people have interfered with this relationship a **plagioclimax** community will result. Grasslands in Europe are mainly plagioclimax communities.

Examiner tip
Know the pattern of temperature and rainfall for a selection of climate types. Also be familiar with the climate of your own home area.

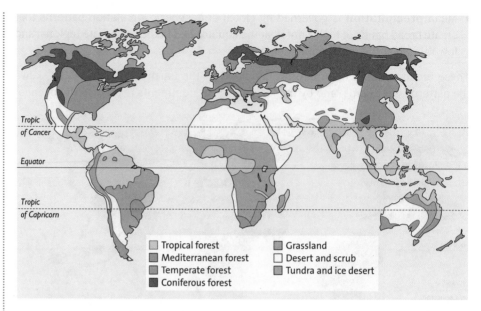

Figure 2 The world distribution of biome types

Every plant species is affected by physical limiting factors and cannot survive beyond its tolerance limits. Water, light and temperature are the most critical factors, although the availability of nutrients is also important. So there is a close resemblance between maps showing the distribution of climate types and the distribution of biomes, as shown in Figure 3.

Examiner tip

Be able to name major countries and regions of the world accurately and to describe distributions effectively. For example, distinguish between different countries and parts of Africa.

Knowledge check 1

What is a biome?

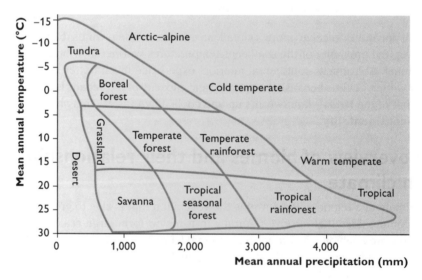

Figure 3 World biome types related to temperature and precipitation

1.2 What are the temporal patterns of climate change?

Climate changes naturally through a series of cycles. Longer-term cycles cause ice ages and interglacial periods when the climate is warmer. Shorter-term cycles may bring periods of warmer climate such as that experienced by people in the Bronze Age around 4,000 years ago or colder climate such as the **Little Ice Age** around 500 years ago.

Short-term climate change

The Little Ice Age occurred between the fifteenth and nineteenth centuries (Figure 4). Between 1607 and 1804, the River Thames regularly froze and people were able to hold fairs on the ice. Evidence for this colder period is found around the world. There were colder winters in Europe and North America, and settlements in the Alps were destroyed by advancing glaciers during the mid-seventeenth century. Average surface temperatures in many regions were at least 1°C lower than those of today.

Examiner tip
Look up historical records of the Little Ice Age.

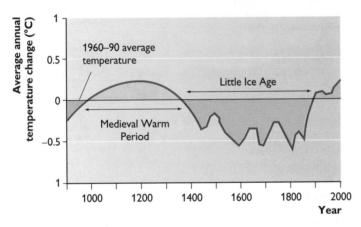

Figure 4 Short-term climate change

The causes of such climate fluctuations are not fully understood, but several climate-change mechanisms have been identified. Individually, these mechanisms cannot explain the Little Ice Age, but a combination of these and other factors will have caused the cooling.

- **Volcanic eruptions** can affect global climate for a year or two by injecting large quantities of sulphur dioxide and dust into the stratosphere. This causes the **aerosol effect** in which incoming solar radiation is reflected, reducing global temperatures by 5–10%. During the Little Ice Age there were numerous volcanic eruptions.
- **Variations in solar energy**, such as the 11-year cycle of sunspots, affect the amount of solar radiation reaching the Earth's atmosphere. During the Little Ice Age, a decrease in observable sunspots was noted. The intensity of solar activity is believed to have fallen by 0.25% — enough for cooling to occur.

- **An increase in the albedo** or reflectivity of the Earth's surface would have caused **positive feedback**. Lower temperatures produced more snow and ice during the Little Ice Age and the lighter surfaces will have reflected more solar energy than darker vegetated surfaces, which absorb energy. This increases the cooling effect, creating the feedback loop.

El Niño and **La Niña** are short-term climatic disruptions that some people think are becoming more frequent and intense because of global warming. El Niño refers to the appearance of warm surface water in the eastern equatorial Pacific, whereas its opposite, La Niña, refers to the appearance of colder-than-average sea-surface temperatures in the central and eastern equatorial Pacific. These events affect climate around the world as they distort **jet stream** flows. Jet streams are high-speed winds that circle the world at high altitudes. Atmospheric connections between widely separated regions are called **teleconnections** and often cause hazardous weather events such as drought or flooding (see Figure 5).

Examiner tip

Up-to-date information can be found on the NOAA web site. In 2011 the site was monitoring a strengthening of El Niño/La Niña for 2011–12. Make sure you know how the effects of El Niño and La Niña differ and where their impacts occur.

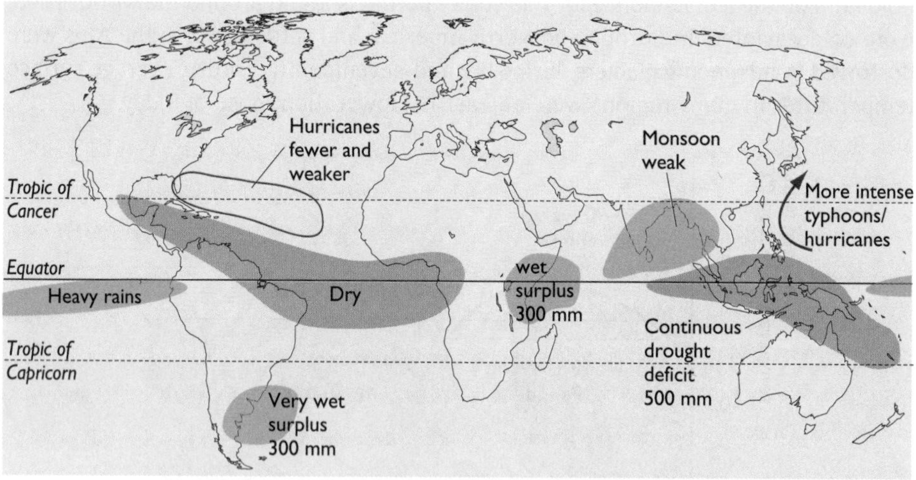

Figure 5 Major impacts of El Niño, June to December 1997

Knowledge check 2

What is the Walker cell?

In a **normal year** (Figure 6), the trade winds (1) blow towards the equator and the warmer water of the western Pacific (4). Here the water heats the air above causing convectional uplift (2), creating low pressure weather conditions and rainfall (6) over southeast Asia (2/3). In the east Pacific off the coast of Peru, the conditions are dry as the air returns and sinks here, creating the **Walker cell** (7). The pressure of the trade winds results in sea levels in Australasia being 50 cm higher than in Peru and sea temperatures being 8°C higher. This air movement also enables the winds to draw up water along the east coast of South America. Here there is a shallow warm upper layer overlying the deep cold layer of the ocean; the boundary between these layers is called the **thermocline**. This causes an up-welling of cold water, bringing nutrients from the ocean bottom and producing optimum fishing conditions off the coast of Peru (5).

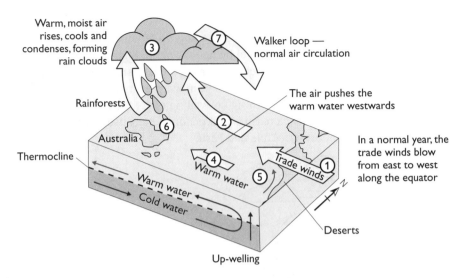

Figure 6 A normal year

In an **El Niño year** (Figure 7), the trade winds (1) in the western Pacific weaken, die or even blow in the opposite direction. The water that is normally piled up in the west moves back towards the east (3) so that the sea level on the Peruvian coast rises by 30 cm and conditions are calmer across the whole Pacific Ocean. The air circulation loop is reversed (2) so that air sinks in the western Pacific, bringing drought to countries in that region, and air rises in the eastern Pacific causing convectional rainfall (4), sometimes leading to floods in coastal areas in South America. The change in the location of rising air distorts the path of the jet streams, causing teleconnections around the world and consequent disruption to normal weather patterns.

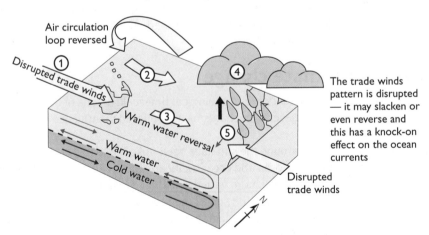

Figure 7 An El Niño year

As the eastern Pacific Ocean becomes 6–8°C warmer, the cold Humboldt Current, which normally flows northward from the Antarctic, is disrupted (5). This changes the environmental conditions upon which the ocean ecosystem depends: phytoplankton cease to grow, which reduces fish numbers, and there is a reduction in food for birds in the Galapagos Islands.

Examiner tip
Practise drawing quick sketches of these diagrams. It is a useful way of conveying a lot of information quickly in the examination.

A **La Niña year** (Figure 8) may be described as an exaggerated normal year, with a very strong Walker loop. There are extremely strong trade winds (1) pushing warm water westwards (2) and raising sea level by up to a metre higher in Indonesia and the Philippines. Deep low pressure develops over southeast Asia, causing heavy rain on account of the strong convectional uplift (3) from warmer-than-usual seas. On the eastern side of the Pacific there is a strong up-welling of cold water off the coast of Peru (4). The higher-than-average pressure that occurs here causes extreme drought in the semi-arid areas of northern Chile and Peru.

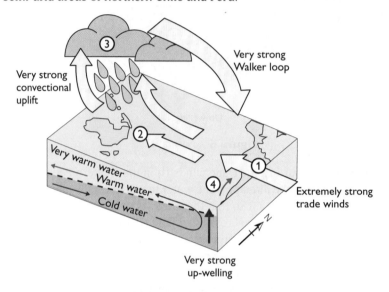

Figure 8 A La Niña year

This see-sawing of atmospheric pressure between the eastern Pacific and the Indo-Australian area is called the **Southern Oscillation** and the full sequence of events is the **El Niño Southern Oscillation (ENSO)**.

Long-term climate change

Long-term climate change occurs over a geological timescale of tens of thousands to millions of years (Figure 9). Several mechanisms are believed to influence the Earth's climate on this scale:

- Shifts in the Earth's orbit around the sun have been linked to longer-term changes. Variations in the Earth's tilt, orbit and eccentricity are known as the **Milankovitch cycles**. A greater tilt produces a more varied seasonal climate. The position of the Earth on its elliptical path around the sun — the precession of the equinox — also varies. There are also changes in the degree of roundness of the Earth's spherical shape, known as **eccentricity**.
- The changing shape and position of the continents on the Earth's surface affects the atmosphere because this affects the **geometry of the oceans** and ocean circulation patterns. The oceans control the transfer of heat and moisture around the Earth and so influence climate. The location of the climate zones alters over millions of years. When the land bridge of Panama was formed 5 million years ago, the flow between the Atlantic and Pacific Oceans was cut off and this may have triggered the ice ages.

Examiner tip

Research online for diagrams illustrating the mechanisms that influence the Earth's climate. This will help you to understand concepts.

- Ocean circulation has changed in response to the changing shape of the continents. Deep ocean currents drive the **thermohaline circulation** that flows between the oceans. This operates like a conveyor belt, transferring heat energy from the tropics to the poles. Changes in ocean circulation have a significant influence on climate as there is a complex relationship between the oceans and the atmosphere. It is possible that a change in weather patterns affected the **ocean–atmosphere conveyor system**, weakening the Gulf Stream or **North Atlantic Drift** that normally brings a warm tropical ocean current to northwest Europe.
- Major volcanic activity, such as the eruption of **flood basalts** that created the Deccan region of India, has been shown to cause global cooling. This may also be linked to the impact of larger asteroids such as at Chicxulub on the Yucatan peninsula in Mexico around the same time. Both have been suggested as causing the mass extinction of 65 million years ago.

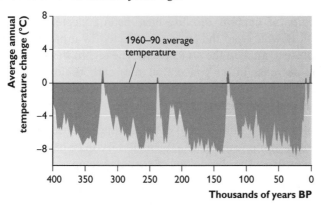

Figure 9 Long-term climate change

The Earth's climate is currently in an interglacial period and some experts believe that another ice age is due. Global warming may be preventing this ice age from occurring. In the latter half of the twentieth century it became clear that the planet was warming. The global mean temperature increased by about 1°C during the twentieth century (Figure 10). The graph shows variability from decade to decade, but the trend is clearly rising. Six of the first ten years of the twenty-first century were the warmest on record. Although the graph shows a levelling off in the last decade, it remains the warmest since reliable records began. Variations from year to year are accounted for by the ENSO; for example, 2008 was a strong La Niña year.

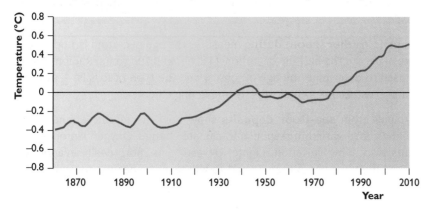

Examiner tip
When studying graphs: read the scales on the axes carefully; read the caption so you understand what the graph is showing; look for trends and patterns; and identify and explain any anomalies. Use data to support your observations.

Knowledge check 3
What is a thermocline?

Figure 10 Average variation in Earth-surface temperature from the mean, 1860–2010

1.3 What are the causes of climate change?

The evidence for climate change

Ice cores drilled from ice sheets in Greenland and the Antarctic contain bubbles that have preserved air over hundreds of thousands of years. From these cores climate scientists have been able to analyse past atmospheric composition and build up detailed records of past levels of carbon dioxide (CO_2). This analysis indicates a positive correlation between temperature change and carbon dioxide concentration in the atmosphere.

Since 1957 records of **carbon dioxide** have been collected at the **Mauna Loa Observatory** in Hawaii. Here the air is not affected by urban-industrial pollution, so it is a good place to analyse the atmosphere. The record shows a seasonal pattern of increase and decrease in the concentration of carbon dioxide — attributable to the greater use of fossil fuels in winter in the northern hemisphere — but the overall trend is upwards and at an increasing rate.

Glacial deposits are a natural record of ice advance and retreat as temperatures have risen or fallen. Moraine and glacial till material can be dated to show the extent of glaciers to reconstruct the chronology of changing temperatures.

Pollen analysis of lake sediments are used to determine past climate. As plants are sensitive to their environment, only certain plant communities will survive in particular climates. By identifying key types of pollen, it is possible to work out what the climate was like in a given area in the past.

Insect species are sensitive to small changes in climate conditions. The remains of beetles (**Coleoptera**) found in lake or land sediments indicate the prevailing climate at the time they were buried. Key species are identified and can be matched to those living today to determine their optimum environmental conditions.

Dendrochronology is the study of tree rings. The bristlecone pine in the high mountains of the Sierra Nevada in western USA is the longest-living tree in the world; some have been recorded as up to 4,600 years old. By matching the rings of fossil trees with living trees, a continuous record of climate over the last 8,000 years has been made.

Radiometric (radiocarbon) dating measures the age of certain isotopes based on their rate of decay. The half-life of carbon-14 is about 5,700 years. Radiometric dating can be used to determine the age of fossils, and the type of fossils — mostly plant matter and crustaceans — indicates the prevailing climate.

Core samples from **sea-floor deposits** show how the marine animal and plant communities have varied in response to climate change. Oxygen isotopes from the sediments also show variation. Lighter oxygen-16 is more easily evaporated and precipitated than heavier oxygen-18 and so will be more heavily concentrated in ice

Examiner tip

Good examination answers include a range of ideas. Draw a spider diagram to help you learn the range of evidence for climate change.

sheets and glaciers during colder periods. This will have left a greater concentration of oxygen-18 in the ocean sediments.

Historical records record the weather conditions. Actual readings are unreliable but accounts of snowfall and frozen water courses or heatwaves and droughts enable a detailed chronology to be assembled. Records of crop yields or prices show how good the weather has been in any given year. Detailed daily weather reports have been kept since 1873 although data have been collected since the seventeenth century.

The atmospheric processes that result in climate change

There is an important distinction between the **greenhouse effect** and **global warming**. In the greenhouse effect, the lower atmosphere is heated as the long-wave radiation emitted from the Earth's surface is absorbed by greenhouse gases. This keeps the average global surface temperatures higher than they would otherwise be. The average temperature on the moon, where there is no atmosphere, is −18°C but on Earth it is +15°C.

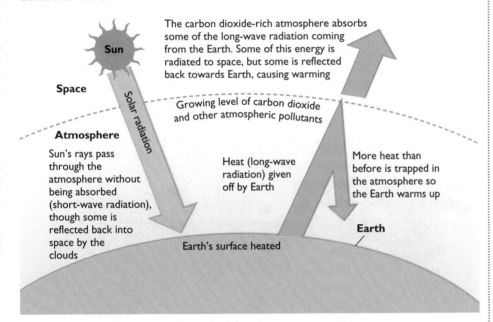

Figure 11 Global warming

Global warming (Figure 11) is the '**enhanced greenhouse effect**', i.e. the effect of an increase in the concentration of greenhouse gases in the Earth's atmosphere. These greenhouse gases are water vapour, carbon dioxide, methane, nitrogen oxides, low-level tropospheric ozone and CFCs. The greater the concentration of these gases, the more the atmosphere will be warmed. Some gases have a more powerful greenhouse effect than others. The enhanced greenhouse effect occurs as the concentration of greenhouse gases increases (Figure 12).

Examiner tip
The distinction between the greenhouse effect and global warming is fundamental to any answer on the current debate about climate change. Always try to show that you know the difference.

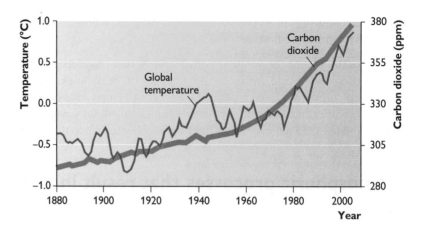

Figure 12 Global temperature and carbon dioxide

The sources of greenhouse gases

Water vapour is a greenhouse gas. Evaporation has increased from manmade water bodies such as reservoirs and irrigation channels, and emissions of steam have increased from power station cooling towers and air conditioning systems.

Carbon dioxide (CO_2) is thought to be responsible for 55–60% of global warming and the acidification of the oceans. Atmospheric carbon dioxide concentration was 270 ppm (parts per million) before the Industrial Revolution. In January 2012 it reached 392 ppm and is continuing to rise at an accelerating rate. Current levels of concentration of carbon dioxide and methane are now the highest in 800,000 years. The increase in carbon dioxide is attributed to the combustion of fossil fuels, deforestation and marine pollution, which kills phytoplankton (plant plankton capable of photosynthesis). Figure 13 shows how carbon dioxide emissions vary globally.

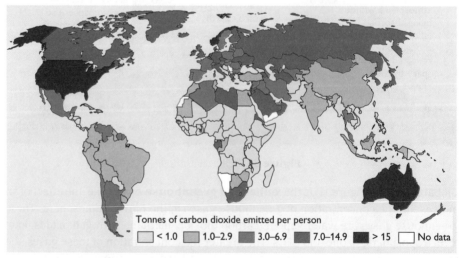

Figure 13 Carbon dioxide emissions per person in 2000

Methane (CH_4) plays a significant role in global warming as it is 25 times more potent than carbon dioxide. In 1750 the atmospheric concentration of methane was

Examiner tip

http://co2now.org/ will provide you with up-to-date information on this greenhouse gas and its effects. Make sure you are up to date with climate change news.

about 700 ppb (parts per billion). In 2010 it was measured at 1,850 ppb. In the past, methane concentrations have ranged between 300 and 400 ppb during glacial periods and between 600 and 700 ppb during interglacial periods. Methane emissions have increased as livestock farming and rice production have expanded to meet the needs of a growing world population. In addition, there are leakages of the gas from natural gas fields and pipelines, coalmines and landfill sites (caused by anaerobic bacteria).

Nitrogen oxides (NO_X) cause about 6% of global warming and their atmospheric concentration has risen from a pre-industrial (*c.*1750) level of 275 ppb to around 315 ppb now. Although their concentration in the atmosphere is much smaller than that of carbon dioxide, nitrogen oxides are more effective greenhouse gases with a global warming potential of 296 over a 100-year time span. This means that 1 kg of nitrogen oxide released into the atmosphere has a global warming effect equivalent to 296 kg of carbon dioxide over a 100-year period. Oxides of nitrogen are emitted from motor vehicles and power stations. Another important source is the production and use of nitrate fertilizers, especially where soils are more water-logged.

Ozone in the stratosphere has been depleted because of the release of CFCs. This is dangerous as ozone plays an essential role in protecting life on Earth by absorbing the most harmful ultraviolet (UV) radiation. However, ozone generated by human activity in the troposphere and CFCs are also greenhouse gases. Low-level ozone is created when sunlight causes a photochemical reaction with nitrogen oxides and hydrocarbons from vehicle emissions.

Chlorofluorocarbons (CFCs) have been used in aerosol propellants, fire extinguishers, refrigerants, air conditioning systems, solvents and expanded plastics (polystyrene). Although their use is being phased out following the Montreal Protocol of 1987, they have a lifespan of over 100 years and so continue to exist in the environment. In the troposphere they act as greenhouse gases and in the stratosphere they destroy ozone.

Table 2 summarises the greenhouse gases and their potential global warming impact.

Table 2 Greenhouse gases

Greenhouse gas (GHG)	Atmospheric concentration (ppm)	Global warming potential (GWP)*	Total potential impact
Carbon dioxide	392	1	392.0
Methane	1.85	25	46.3
Nitrogen oxides	0.315	296	93.2
Ozone	0.04	2,000	80.0
CFCs	0.0004	14,100	5.6

* Global warming potential (GWP) is an estimated measure of the contribution made by each gas to global warming over a given time period (in this case, 100 years). The mass of each greenhouse gas (GHG) is compared to the same mass of carbon dioxide, whose GWP is equal to **1**.

The evidence for global warming

The most compelling evidence for global warming is that mean global temperature is rising. It is argued by some that the rise is small — 1°C in a 100 years — and that the

Knowledge check 4

What are the main GHGs and what have been the trends in their concentrations in recent years?

Knowledge check 5

Distinguish between the long- and short-term causes of climate change.

Earth's climate has always fluctuated. The increase seems to have slowed in the first decade of the twenty-first century, suggesting that the rise may not be a trend but just another cycle. Temperatures in the upper atmosphere are cooling, but this may be connected to the warming of the lower atmosphere.

The **retreat of ice sheets and glaciers**, and **rising sea levels**, may be entirely natural responses to fluctuating temperatures. These trends could easily reverse if the climate cooled again. However, in the last 30 years, most glaciers and ice sheets in the world have shrunk and sea levels are rising.

Following the Little Ice Age, glaciers retreated until about 1940 as the climate warmed. Then glacial retreat slowed and even reversed in some places between 1950 and 1980 as a slight global cooling occurred. However, since 1980 glacial retreat has become increasingly rapid. With few exceptions all the glaciers of the world are shrinking; since 1960 they have lost an estimated 8,000 km³ of ice. In the first 5 years of the twenty-first century, virtually all the glaciers in the Alps were recorded as retreating and at a faster rate than in previous decades.

In the past 125 years, the Athabasca Glacier in the Canadian Rockies has lost half of its volume and receded more than 1.5 km. All 47 of the glaciers monitored in the Cascade Range of western North America are receding and five have disappeared completely. This pattern is repeated around the world — in South America, New Zealand and the Himalayas. Glaciers in the Mount Everest region of the Himalayas are retreating rapidly. The Rongbuk Glacier, which drains the north side of Mount Everest into Tibet, has been retreating 20 m per year and in the Khumbu region of Nepal the average retreat was 28 m per year. There is almost no ice left on Mount Kilimanjaro, the highest mountain on the African continent. Figure 14 shows the loss of glacier ice since 1980.

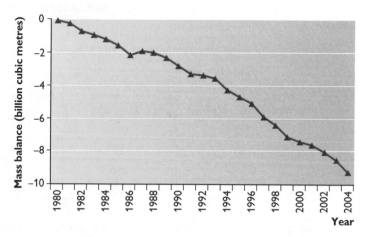

Knowledge check 6

Summarise the information given in Figure 14.

Figure 14 Global cumulative glacier mass balance

The largest glacier in France, the **Mer de Glace**, lies in the Chamonix valley on the northern slopes of the Mont Blanc massif. In the eighteenth and nineteenth centuries the glacier descended to the hamlet of Les Bois. Its channel is 11 km long and 400 m wide but in the last 130 years it has lost 8.3% of its length (about 1 km) and narrowed by 27% (150 m). The glacier moves about 90 m downslope each year, equivalent to about 1 cm an hour. It originates at an elevation of 2,400 m and was fed by the confluence of two glaciers, du Géant and de Lechaud, and the Cascade du Talèfre, but now only the Glacier du Géant flows to the confluence (Figure 15). Glacial meltwater lakes have formed at the foot of the Mer de Glace, trapped behind moraine dams. If these dams are breached, the valley below could be flooded.

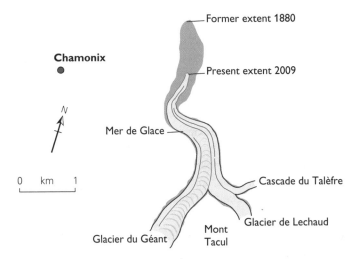

Figure 15 The extent of the Mer de Glace near Chamonix, France

The mountain and valley glaciers form a small fraction of the ice on the Earth. Most is held in the ice sheets of Antarctica and Greenland, up to 3 km or more deep, which feed many outlet glaciers carrying ice to the surrounding oceans. Recession is occurring in these outlet glaciers and this is destabilising the ice sheets. Three such glaciers — Helheim, Jakobshavns and Kangerdlugssuaq — together drain nearly a fifth of the **Greenland ice sheet**. These glaciers have **calving fronts**, which means their snouts reach out to sea and break off to form icebergs.

- Satellite images and aerial photographs have been used to study the movement of the Helheim Glacier. These show that it was stable in the mid-twentieth century. However, at the beginning of the twenty-first century it began to retreat rapidly, receding by 7.2 km in just 4 years and at an accelerating rate from 20 to 32 m per day.
- Jakobshavns in west Greenland is probably the fastest-moving glacier in the world at speeds of over 24 m per day with a 12 km-long floating terminus that was stable for many years. However, in 2002 it began to retreat rapidly, the ice front started to break up and the terminus disintegrated as its movement accelerated to over 30 m per day.

The rapid melting of these glaciers is a clear sign of climate change. It is attributed to rising sea temperatures as the acceleration is most pronounced at their fronts and impacts on the ice sheet feeding them.

Mean sea level (MSL) is a measure of the average height of the ocean's surface; it is used to measure land elevation. A change in sea level may occur because of a real change in sea level or because of a change in the height of the land. A change in sea level relative to the land is called **eustatic change**, whereas a change in the

Examiner tip
The United Nations Environment Programme (UNEP) publishes data on glacial retreat by region: **www.grid.unep.ch/ glaciers/**. You should obtain some up-to-date figures to impress the examiner.

Knowledge check 7

Distinguish between eustatic and isostatic change.

height of the land is an **isostatic change**. Regions covered in ice sheets during the last ice age were depressed by the weight of the ice and are still rising or recoiling following the retreat of the ice. In Britain, mean sea level is measured by tide gauges to establish zero metres on Ordnance Survey maps. Measures taken at 23 stable tidal sites around the world show that mean sea levels rose by 20 cm during the twentieth century. Since 1992 satellite altimeters have made more precise measurements of sea level. The eustatic rise in sea level is due to the thermal expansion of ocean water and the melting of ice sheets and glaciers on the land. The melting of sea ice does not add to sea-level rise; icebergs, for example, displace seven-eighths of their area in the water and water contracts when it changes from frozen to liquid states.

Frozen soil underlies much of the Arctic and sub-Arctic. While the deeper permafrost layer remains frozen, the upper layer thaws and freezes with seasonal temperature change. As the climate warms, the **permafrost layer is melting** to a deeper level. This has led to building and road damage, shrinking lakes as water percolates to aquifers below, rapidly eroding river banks causing increased sediment transfer, and the disappearance of some wildlife, including fish. The local weather may also be affected by the loss of lakes because there will be less water evaporated to form clouds and rain.

Extreme weather events seem to be increasing and this is linked to global warming. As air temperature rises, so do ocean temperatures. Storms draw their energy from the heat of the oceans so this could increase the intensity and frequency of hurricanes. Droughts have also increased in terms of length and intensity, affecting millions of people in Spain, Australia and east Africa.

The **boundaries between the biomes are moving**. Forests of spruce trees in the Yukon in northern Canada are taking over former tundra landscapes and forcing out the species that live there; caribou, for example, have declined across south-western Yukon. Scientists have found that the tree line is moving northward, and higher up mountains, more rapidly than expected. It advanced considerably during the latter half of the twentieth century as tree density increased by up to 65% on cooler, north-facing slopes, and by as much as 85 m higher on warm, south-facing slopes. Changes in the vegetation cover alter the albedo as coniferous trees absorb more sunlight than the tundra. This energy is then re-emitted to the atmosphere as heat, creating a positive feedback mechanism that adds to global warming.

The relative role of environmental and human factors in recent climate change

Increasingly, scientists agree that human activities are causing global warming, but climate science does not present certainties. Large quantities of data are collected and computed and detailed models are developed in an attempt to match observed with predicted outcomes. However, the ocean–atmosphere system is immensely complex and scientists still have much to learn about its mechanisms.

Knowledge check 8

What is the difference between natural and anthropogenic forcing?

Climatologists have attempted to separate natural causes of climate change from human or anthropogenic (i.e. generated by people) causes. **Natural forcing** does not completely explain the steady rise in global temperature that has occurred in the last 50 years, whereas graphs of predicted temperatures based on **anthropogenic** forcing show a closer match. When the two are combined there is a close fit between the predicted temperatures and those actually observed. This provides compelling evidence that people are the cause of global warming (Figure 16c).

According to the **'Ruddiman hypothesis'**, humans have been changing the climate for about 8,000 years as people developed intensive farming techniques. He claims that the arrival of the next ice age has been held up by the activities of early farmers, which increased carbon dioxide and methane emissions. He notes that periodic plagues, which reduced human populations, led to natural reforestation and these episodes can be matched to dips in greenhouse gases.

On the other hand, scientists studying the effects of solar radiation on Arctic temperatures have found a much stronger correlation between temperature and **solar irradiance** (the amount of electromagnetic energy being received on a surface per unit time per unit area) than with carbon dioxide. However, there are few temperature recording stations in the Arctic region and data for global climate warming come from an average of data from many more stations around the world.

Another school of thought suggests that the Earth's planetary self-regulation will occur. The **Gaia principle**, proposed by James Lovelock, suggests that the physical components of the Earth form a complex interacting system of inter-relationships between the atmosphere, oceans, biosphere and lithosphere. He suggested that **negative feedback** maintained a stable state within these spheres, preventing large changes in temperature, for example. Some climate-change sceptics suggest that the Earth can cope with whatever pollution people create, but Lovelock is far more pessimistic. He believes that damage to the biosphere and the increase in greenhouse gas emissions is testing the Earth's ability to self-regulate, increasing the chance of **positive feedback** and leading to runaway global warming.

Although climate-change sceptics point to errors and misjudgements in climate science, the overwhelming majority of scientists agree that there is a 90% chance that people have influenced climate change in the last 50 years.

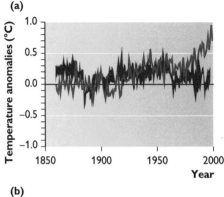

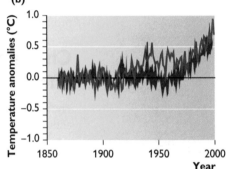

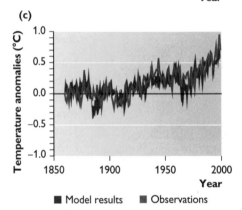

Figure 16 Comparisons between modelled and observed temperature rise, 1860–2000: (a) natural forcing only, (b) anthropogenic forcing only, (c) natural and anthropogenic forcing

1.4 What are the issues resulting from climate change?

The key facts about climate change are:
• Average global surface temperatures increased by 0.6°C during the twentieth century.

- Global warming cannot be explained by natural factors alone. Emissions of greenhouse gases from human activities are needed to explain the rise in temperature.
- Current climate models predict that global temperatures will rise by a further 1.4–5.8°C by the end of the twenty-first century.
- The observed rise in sea level during the twentieth century may be related to this increase in global mean temperatures.
- Global mean sea levels are predicted to rise by between 79 and 88 cm by 2100.

Despite policies to limit greenhouse gas emissions, it is expected that the temperature will continue to rise. This will have far-reaching consequences on all aspects of the global environment, economy and society. Climate change will affect the way of life for millions of people as it will impact on water supply, food production, health and property.

Changing climates, shifting climatic belts and the effect on biomes

The **migration of climate belts** will impact on hydrology, ecosystems and economic activities. Temperate grasslands on the steppes of north Asia and the North American prairies will have drier summers and colder winters. These regions are the 'bread baskets' of the world — the principal grain-growing areas — so climate change will reduce grain yields, food production and income for farmers. It may reduce world food production, possibly leading to food shortage and famine. However, creeping desertification in the Sahel may be arrested as the rain belt moves north.

There is growing evidence of the **ecological impacts** of recent climate change in all biomes. Plant and animal life are responding to changes in air temperature, soil moisture and water salinity. Every species has particular optimum environmental conditions for survival so, as climate alters, they must adapt or migrate to survive. Therefore, plants, animals and insects are invading new areas and mixing with formerly non-overlapping species. This impacts on their environment, causing knock-on effects throughout the ecosystem and sometimes causing extinctions as the structure of the ecosystem is altered.

Two types of ecological impact have been noted in the USA: shifts in **species' ranges** (the locations in which they can survive and reproduce) and shifts in **phenology** (the timing of biological activities that take place seasonally). About 40% of wild plants and animals studied are relocating to stay within their tolerable climate ranges. Those that cannot move fast enough, or whose ranges are shrinking, are unlikely to survive. As Arctic sea ice shrinks, the habitats of polar bears and seals are disappearing. Climate change is driving changes in the timing of seasonal behaviour: species respond to spring 2 to 3 weeks sooner than just a few decades ago. Migrant birds are arriving earlier, butterflies are emerging sooner and plants are growing earlier.

The northward movement of **butterfly distributions** is also being seen in Europe. Over 60 species of butterfly with widespread food-plants are known to have spread north in Europe and other species have moved further up mountains. For example, the *Polygonia comma* is spreading north in the UK at a rate of 10 km per year. Moths are good indicator species because they are sensitive to environmental changes. They are spreading northward and new species are arriving from mainland Europe, having migrated over 150 km north in the past 30 years.

Pests and diseases such as malaria, will also move beyond their usual locations as environments change, affecting greater numbers of people and agricultural enterprises. Malaria is one of the biggest killers in tropical and subtropical climates. Children are particularly vulnerable and it is a major hindrance to economic development. Malaria has been eradicated from Australia but it could return to Queensland if current trends continue.

Warming may alter the location of the **tundra** and the **northern tree-line**, changing the way of life for people living in these regions. The severing of longstanding ties to the land that has sustained the Alaskan native peoples for hundreds of generations would adversely affect their culture and way of life. **Melting permafrost** affects the habitats, wildlife and people who rely on the land for their livelihood. It causes subsidence and damage to settlements, transport and oil pipelines in Alaska and Siberia.

Melting permafrost along river banks, such as the Yukon River and its tributaries in Alaska, increases **erosion**. Villages located along rivers are threatened as the permafrost no longer keeps the river banks firm. Sedimentary material is released into the water, making the rivers shallower and affecting both animal and human populations including the salmon that migrate upstream to spawn.

Organic material and gases have been trapped within the permafrost in soils that have been frozen for thousands of years. Melting permafrost may accelerate global warming by **releasing methane and carbon dioxide** into the atmosphere. This sets up another positive feedback loop as global warming is accelerated. Most of the methane-releasing permafrost is in northern and eastern Siberia. The amount of carbon trapped here may be 100 times greater than the amount of carbon released into the air each year by burning fossil fuels and it is escaping from the thawing permafrost at a much faster rate than expected. Methane is also stored in the seabed as methane gas or methane hydrates and then released as sub-sea permafrost thaws. The Arctic Ocean sea floor holds vast stores of frozen methane. It is showing signs of instability and widespread release of this greenhouse gas.

Knowledge check 9

What do you understand by positive and negative feedback?

Increasing levels of extreme weather and the impacts on human activities

Extreme weather is predicted to become more common as climate change accelerates with continued global warming. Some climate scientists believe that this is already happening and that climate change will be quicker and more severe than previously expected. In fact, extreme events occur when natural variations in the weather and climate combine with long-term climate change. Directly linking any one specific extreme event to global warming is not possible, but climate change may increase the probability of weather events becoming more extreme.

People's lives may be put at risk from an increased frequency of droughts and flooding. Some parts of the world will receive more rain while others, especially existing deserts, may receive less than usual on account of changes in the hydrological cycle and the intensification of weather patterns (Figure 17).

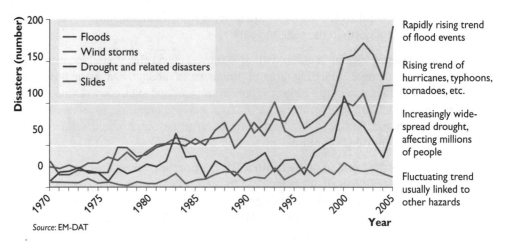

Source: EM-DAT

Figure 17 Hydro-meteorological disasters, 1970–2010

The number of regions affected by **drought** has increased because precipitation has marginally decreased while evaporation has increased because of warmer conditions. In the Murray-Darling basin in Australia, a severe drought prevailed through the first decade of the twenty-first century and the country will need several years of above average rainfall to recover. A number of factors combined to make the drought dangerous: reduced rainfall, increased temperatures and increased population. The wetlands are being seriously degraded, with 50–80% severely damaged or completely destroyed. Agricultural output was reduced by over 60%.

In some areas water resources for drinking and irrigation will be affected by reduced rainfall. An additional 3 billion people could suffer increased water stress by 2080; North Africa, the Middle East and the Indian subcontinent will be the worst affected. Loss of tropical rainforests in northern Brazil and central southern Africa — on account of lower rainfall and higher temperatures — could reduce the effectiveness of this major **carbon sink**, further adding to the increase in carbon dioxide in the atmosphere.

It has been predicted that the frequency of **heatwaves** will increase. Since 1950 the number of heatwaves has risen and there have been widespread increases in the numbers of warm nights recorded. An additional 40,000 deaths in western Europe were attributed to the heatwave of August 2003. If heatwaves are combined with extended dry periods, forest fires increase and become more intense. Victoria, Australia in February 2009 saw some of the worst bushfire-weather conditions ever recorded. Temperatures of 43–48°C and wind speeds in excess of 100kph during an intense heatwave created several large firestorms, particularly northeast of Melbourne where a single firestorm caused 120 deaths.

The number of **heavy daily precipitation** events causing flash flooding has risen as higher temperatures increase evaporation and cloud formation. The poorest countries are the most vulnerable to the effects of climate change: 80 million people are thought to be at risk of **flooding**, mainly those living in south and southeast Asia (such as the 2011 floods in Thailand). Heavy rainfall in the UK— leading to major flood events such as Tewkesbury, Hull and Sheffield in 2007 and Cockermouth in

2009 — has also been linked to global warming, but the increasing frequency of such events may be the result of mismanagement of water channels and increased building on floodplains.

The **North Atlantic Oscillation (NAO)** is the fluctuation of atmospheric pressure at sea level in the North Atlantic Ocean between the Icelandic low pressure and the Azores high pressure. This east–west oscillation controls the strength and direction of westerly winds and storm tracks across the North Atlantic. This is predicted to decrease as the Arctic becomes warmer, reducing the number of storms in the mid-latitudes.

There is no clear evidence of a long-term trend in **tropical storm** activity despite several severe hurricanes in the Caribbean and cyclones in the Bay of Bengal in recent years. However, there is a link between sea-surface temperatures and the intensity of storms, which seems to be related to the occurrence of **El Niño** and its teleconnections. El Niño events increase tropical cyclone activity around the Pacific and decrease it in the Atlantic, whereas **La Niña** events bring opposite conditions. Climate change has been linked to an increase in the frequency and intensity of El Niño events, increasing the chance of extreme weather events.

Economic damage as a result of extreme weather events has dramatically increased over recent decades. This is partly on account of population growth and the rise in global wealth held in property increasing the costs of extreme weather events. However, the insurance firm Munich Re compared losses in the 1960s and in the 1990s and found that a major part of the increased losses was because of changes in the frequency of extreme weather events.

Melting glaciers, rising sea levels and their impacts on people

The **retreat of glaciers** affects the availability of fresh water for irrigation and domestic use, mountain recreation such as skiing, the animals and plants that depend on glacier melt, and the level of the oceans in the longer term.

Glaciers act as stores of water to maintain the flow of rivers. An estimated 1.5 to 2 billion people in Asia, Europe and North and South America depend on glacial meltwater to supply their catchment areas. Increased ablation (glacial melting) will increase river runoff initially, but then there will be a sharp decline. Reduction in runoff will affect the ability to irrigate crops and will reduce the river flows needed to keep dams and reservoirs replenished for domestic and industrial use.

All coastal processes are affected by sea-level changes. In the past, eustatic change created rias (drowned river valleys) and fjords. Sea levels are expected to rise because of the thermal expansion of the water as its temperature rises and because of melting glaciers and ice sheets on the land (not sea ice as this displaces its own liquid volume in the ocean). There will be significant **coastal impacts** with coastal erosion, flooding and the possible loss of coastal wetlands and their associated flora and fauna, especially wading birds and shallow-water fish species. Salt water will also pollute groundwater supplies as it percolates through permeable rocks along the coastline, as has happened in the Perth area of Western Australia. If sea levels

Examiner tip

Research the latest extreme weather events on the World Meteorological Organization's website. You can keep up to date with hazardous weather around the world as it happens — the news websites from the BBC and *Guardian* are useful.

Knowledge check 10

Distinguish between **El Niño** and **La Niña**.

Examiner tip

Find some examples of the impact of glacial melting and sea level rise from the IPCC publication *Sea Level Rise and Ice Sheet Instabilities* (2010). You will find it on their website. For top-level answers you need to not only say that low-lying land will be affected, but to suggest where and by how much.

continue to rise (Figure 18), major deltas could disappear and low-lying coastal areas of the Netherlands and the UK (such as the Fens and Dungeness) will be severely eroded; high-value property and important industrial and commercial developments are seriously at risk of flooding. The areas most vulnerable to sea-level rise are low-lying and heavily populated places such as Bangladesh and ecological hotspots such as coral reefs in Tuvalu and Aitutaki.

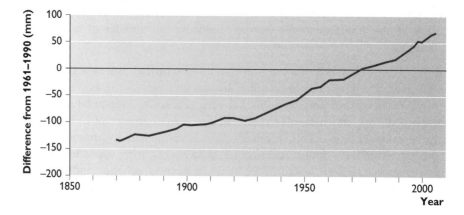

Figure 18 Global average sea level, 1870–2005

The Maldive Islands are a series of coral atolls in the Indian Ocean rising no more than 1.8 m above sea level. If sea levels rise by over 1 m by 2100, as predicted, 385,000 people will have to evacuate the islands. In May 2007 seasonal high tides, exacerbated by intense winds, flooded 55 of the 194 inhabited islands, damaging homes, crops and infrastructure. Malé, the capital, is surrounded by a 3 m-high wall that protects against the fortnightly tidal surges but not rising sea levels.

The island of Lohachara in the Sundarbans in India was the first inhabited place to be abandoned to the sea. The Sundarbans is one of the largest **mangrove forests** in the world, located on the delta of the Ganges, Brahmaputra and Meghna rivers in the Bay of Bengal. The sea level here has risen by over 3 cm per year in the past 20 years and a rise of 1 m in sea level would inundate 1,000 km². Four islands have so far been lost to the sea, making 6,000 families homeless. The entire island system is facing rapid coastal erosion, flooding and salinisation of drinking water. By 2020 the rising sea level and soil erosion threaten to displace more than 30,000 people. The region has a diverse marine life which, if harmed, will drastically affect the food chain and the fishing industry. On account of its exceptional biodiversity, including 260 bird species and the Bengal tiger, it is designated a World Heritage Site. The forest is South Asia's largest **carbon sink**, absorbing carbon dioxide and helping to counter global warming.

The variation of these impacts in different regions

Equatorial heat is moved towards the poles by huge air circulation cells and by **ocean currents** that are controlled by temperature and salinity. Melting ice reduces ocean salinity and so may change the pattern of ocean currents. The North Atlantic Drift carries warmer water from the Caribbean Gulf Stream, creating a climate in northwest Europe that is some 8°C warmer than the mean value at this latitude. This system may be especially sensitive to climate change and already shows some signs of diminishing. Significant change in the temperature and location of ocean currents could have a major impact on climate and weather systems. The most extreme scenario envisages a rapid change in the ocean conveyor belt that circulates water around the world.

Coral reefs act as barriers to wave action by dissipating most of the energy before waves reach the beach, but they are particularly vulnerable to climate change. Tropical corals grow in water temperatures of between 20°C and 30°C and depths of 40 m or less. It is a very fragile ecosystem, so sea-level rise and **warmer sea temperatures** threaten their existence, causing bleaching and death. Rising sea levels may also outstrip the corals' rate of growth, submerging them below their preferred depth.

Ocean acidification causes serious harm to marine organisms such as corals, lobsters and sea urchins. Although this moderates the increase of carbon dioxide in the atmosphere, when carbon dioxide dissolves in sea water it forms carbonic acid and acidifies the ocean. Fixed organisms such as corals cannot survive; coral bleaching is already affecting fragile reef ecosystems. Other species migrate; for example, the sea bass, a Mediterranean species, is now being found around the coasts of Britain.

How the UK could be affected

New climate change scenarios were launched by the Department for Environment, Food and Rural Affairs (Defra) in April 2002 based on a rise in temperature of between 2°C and 3.5°C by the 2080s. It is expected that there will be greater warming in the south and east than in the north and west of the UK and that there will be more frequent hot summers and fewer cold winters. It is also thought that winters will become wetter with less snow and summers drier. The southeast in particular could see a decline in summer rainfall of up to 50%.

Many homes and businesses, and coastal habitats for birds and other wildlife, are now under threat of flooding and erosion with rising sea levels and more intense storms. Sea levels are expected to rise by 26–86 cm above the current level in southeast England by the 2080s. There is an increasing risk of a damaging storm surge such as the one in 1953 that flooded Canvey Island and threatened much of central London. Extremely high water levels could become 10 to 20 times more frequent by the 2080s.

Weather-related impacts on people's quality of life, the economy and the natural environment also include:
- heavier rainfall increasing pressure on drainage systems
- summer drought causing low stream flows, water shortages and the risk of subsidence
- heatwaves bringing health risks, especially to older people

Most of the advantages of climate change relate to farming, agriculture and harvesting due to a warmer climate. Cereal crops start growing when the temperature rises above 6°C, so the UK is seeing an extended growing season. Timber yields have increased as the tree line has moved from the 500 m to the 550 m contour. The growing period for commercial species of pine and spruce has reduced by about 10 years. There are

Knowledge check 11

What is a carbon sink?

Examiner tip
You can learn about ocean acidification by using the *IPCC Workshop on Impacts of Ocean Acidification on Marine Biology and Ecosystems* (2011). Use of the latest publications will help you to keep up to date.

nearly 400 commercial vineyards, all in the south of England and Wales, covering approximately 1,324 hectares (ha) of land benefiting from warmer temperatures and fewer frosts. Other potential benefits include being able to produce a wider variety and more food in the UK, which might reduce reliance on imported foodstuffs, the expansion of tourism in some areas and lower costs for heating, although more air-conditioning may be needed in summer.

The impacts of climate change on society

Changes in the nature and frequency of extreme weather events will have social impacts. Economic development, education, healthcare and public-health infra-structure are important factors in enabling societies to adapt to climate change and to protect people from disease and injury as a result of it. Economic damage, social disruption and loss of life could be substantial. Climate change may mean a global redistribution of the costs and benefits of the weather.

Food supply

Climate change will affect agriculture and food production around the world because of:
- the effects of more carbon dioxide in the atmosphere, which enhances photosynthesis
- higher temperatures
- changes in the pattern of precipitation and transpiration
- increased frequency of extreme events
- the spread of pests and diseases

Africa is expected to experience significant reductions in cereal yields, as are the Middle East and India, whereas regions at higher latitudes will have increased crop yields.

Health

It has been found that climate change has altered the seasonal distribution of some allergenic pollen species and the distribution of some infectious disease vectors. It has also increased heatwave-related deaths. It is suggested that there may be:
- increased malnutrition
- increased deaths, diseases and injury due to extreme weather events
- increased occurrence of diarrhoeal diseases
- increased frequency of cardio-respiratory diseases
- altered spatial distribution of some infectious diseases

An additional 290 million people could be exposed to malaria by the 2080s, with China and central Asia likely to see the largest increase in risk. However, climate change could bring some benefits in temperate areas such as fewer deaths from cold exposure.

Water resources

The negative impacts of climate change on water resources and freshwater systems are expected to outweigh the benefits. Semi-arid and arid areas are particularly exposed to these impacts because many of these areas — such as the Mediterranean basin, western USA, southern Africa and northeastern Brazil — would suffer a decrease in water supply. This will impact on both food supply and health.

Migration and conflict

There is likely to be competition over natural resources that are increasingly scarce as a result of climate change. Problems arising from increased drought, water shortages and river and coastal flooding would affect many people, leading to relocation within or between countries. This might exacerbate migration pressures and ethnic conflicts, causing political tension.

1.5 What strategies can be used to address climate change?

The UK government commissioned the **Stern Review** on the Economics of Climate Change, published in 2006, which assessed evidence for the impacts of climate change and for the economic costs. It concluded that the benefits of strong and early action far outweigh the economic costs of not taking action.

- The overall costs and risks of climate change will be equivalent to losing at least 5% of global GDP each year if no action is taken and could rise to 20% of GDP or more.
- The costs of action — reducing greenhouse gas emissions to avoid the worst impacts of climate change — could be limited to around 1% of global GDP each year.

Stern recommended four key elements for future international frameworks:

- **Emissions trading** to promote cost-effective reductions in emissions and to bring forward action in developing countries to support the transition to low-carbon development paths.
- **Technology cooperation** to boost the effectiveness of investments in innovation, supporting energy research and development and the use of new low-carbon technologies to improve energy efficiency.
- **Action to reduce deforestation** to reduce the loss of natural forests around the world, which contributes more to global emissions than transportation.
- **Adaptation** by fully integrating climate change into development policy, ensuring that rich countries honour their pledges to increase development assistance and researching new crop varieties that will be more resilient to drought and flood because the poorest countries are most vulnerable to climate change.

There are two ways to address climate change: mitigation and adaptation.

Mitigation strategies

Mitigation involves reducing greenhouse gas emissions and increasing the sinks for these gases. This can be done by setting targets to reduce emissions, switching to renewable energy sources, and carbon capture and storage.

Examiner tip

When you are researching climate change, always try to find who has written the document and whether they may be biased. Because there is no absolute certainty about the future, views may be prejudiced, especially if the comments are from a political party or an interest group. Showing awareness of this may gain marks in the exam.

Replacing fossil fuel consumption is the obvious mitigation strategy but, as a new coal-fired power station comes on stream in China every week, it is probably an unrealistic option until coal resources approach exhaustion. Fossil fuel use is increasing globally as people in developing countries aspire to the same living standards as those in advanced nations. Motor vehicles continue to be fuelled with petrol or diesel as this is the more efficient option, although dual-fuel, hybrid and electric alternatives are coming on to the market.

Viable **renewable energy options** for electricity generation include solar power, wind turbines, tidal and wave energy, HEP, geothermal energy and the use of biofuels. All have their advantages and disadvantages and together they could contribute a substantial portion of energy supplies, but their development will involve considerable investment in new infrastructure.

Energy conservation is an attractive option as it also saves money for the consumer. People are being encouraged to use low-energy light bulbs, insulate their homes properly, avoid leaving appliances on standby and drive in a more fuel-efficient way. Developers are being urged to build zero-carbon buildings in both the public and private sectors. The Green Deal introduced in 2010 is an extension of these policies.

Local authorities are instigating a range of **transport policies** that reduce congestion and pollution as well as saving energy. These include congestion charging, light railway and tram schemes, increasing car park charges and encouraging cycling and walking.

The **forestry** sector can help to tackle climate change by increasing the carbon sink. Limiting deforestation and extending the conservation of existing forests, restoring forest cover and creating new forests such as the National Forest in the English Midlands will offset some carbon emissions.

Carbon offsetting is a financial mechanism to reduce emissions. One carbon offset represents the reduction of one tonne of carbon dioxide or its equivalent in other greenhouse gases. There are two markets for this:
- In the larger, compliance market, companies and governments can buy carbon offsets to enable compliance with their agreed caps on carbon emissions.
- In the smaller, voluntary market, individuals, companies or governments can buy carbon offsets to mitigate their own greenhouse gas emissions from transportation, electricity use and other sources.

Knowledge check 12

Give one example for each of the mitigation strategies.

Offsets are made through financial support of green projects such as renewable energy developments, forestry schemes or the elimination of sources of pollution.

Adaptation strategies

Adaptation involves changing lifestyles to cope with the consequences of climate change. This includes managed retreat from eroding coastlines, the development of drought-resistant crops and the extension of conservation zones to enable the migration of species.

International level

On 9 May 1992 the world's governments adopted the **UN Framework Convention on Climate Change** at the Earth Summit in Rio de Janeiro. The aim was to consider action to reduce global warming and to cope with the consequences of temperature increases. This led to the **Kyoto Protocol**, an international agreement on fighting climate change, which was agreed in Japan in December 1997 with the objective of reducing the greenhouse gases to their 1990 levels in the years 2008–12 and became law in February 2005. The USA agreed to a 7% reduction in greenhouse gases, the EU to an 8% reduction, Japan to a 6% reduction and 21 other countries agreed to similar targets. However, the USA's target is not binding as it has so far declined to ratify the agreement.

The **Intergovernmental Panel on Climate Change** (IPCC) was set up in 1989 by the United Nations Environment Programme and the World Meteorological Organization to provide a well-researched scientific view on the current state of climate change and its potential environmental and socioeconomic consequences. The IPCC's first report in 1990 led to the creation of the UN Framework Convention on Climate Change — the key international treaty to reduce global warming and cope with the consequences of climate change. Its fourth report was published in 2007.

Examiner tip

Assess the international efforts to respond to climate change by looking at the IPCC *Special Report on Managing the Risks of Extreme Events and Disasters to Advance Climate Change Adaptation* (2011). You can also search the news media websites such as the BBC.

National government action

The **Climate Change Act 2008** made Britain the first country in the world to set legally binding carbon budgets with the aim of cutting UK emissions by at least 80% by 2050.

The act set greenhouse gas emission targets in legislation and established a system of 5-year carbon budgets. These became law in May 2009 and require reductions to below 1990 levels of:
- 22% in 2008–12
- 28% in 2013–17
- 34% in 2018–22

The **Low Carbon Transition Plan** (LCTP) outlines how the UK will achieve these carbon budgets and save around 700 million tonnes of carbon dioxide equivalent by 2022. Government departments were given a share of the UK's carbon budget for which they have responsibility to reduce emissions. In March 2010 every department produced a Carbon Reduction Delivery Plan setting out in detail which actions it will take. The Department of Energy and Climate Change (DECC) oversees these plans and monitors delivery. The Act also sets out a framework to plan for future climate risks and for the government to:
- work with the private sector to improve awareness of climate risks and to encourage new business innovation
- protect national infrastructure from climate risks
- take action to safeguard the natural environment
- ensure that cities are designed and built, or adapted, to cope with the challenges of pollution and the urban heat island effect
- improve heatwave plans to reduce the expected health impacts of climate change
- understand the risks and opportunities that the UK faces as a result of climate impacts in other countries

In 2004 carbon emissions per person in Wales were the highest in the UK and the twelfth highest in the world, producing 14.2 tonnes of carbon dioxide per person compared to 8.8 tonnes in England. The Welsh Assembly has taken steps to reduce the country's emissions, including:

- aiming to achieve annual carbon emission reductions of 3% per year by 2011 in all devolved areas
- setting up a Climate Commission for Wales
- developing schemes to install renewable microgeneration technology in fuel-poor households and to support community energy generation
- publishing the Renewable Energy Route Map, which proposed that, within 20 years, Wales can produce as much electricity from renewable sources as it consumes
- proposing that, from 2011, all new buildings in Wales should be constructed to zero carbon standards

The **Carbon Trust** was set up in 2001 by the UK government as an independent not-for-profit company with the mission to accelerate the move to a low-carbon economy. It works with other organisations by providing specialist support to reduce carbon emissions today and to develop commercially viable low-carbon technologies that will reduce carbon emissions in the future. It aims to save over 20 million tonnes of carbon a year by 2050.

Local policies

The **Keep Leicester Cool** campaign promoted ten steps that people can take to reduce their carbon dioxide emissions. It was set up in 2003 to provide practical solutions to encourage individuals to take personal responsibility and reduce the city's impact on climate change. The 3-year campaign, using both print and broadcast media, focused on the issues of flooding, home energy, local produce, health, renewable energy, gardening, recycling, transport and saving water. The scheme specifically targeted Asian audiences, who make up over 30% of the local population.

Launched in 2003, **CarbonNeutral Newcastle** is a city-wide initiative to raise awareness of climate change and reduce the carbon emissions of people and businesses. Its objectives include raising the profile of climate change, helping people to reduce their carbon emissions, offering help to businesses and raising funds for low-carbon projects.

Swansea's **Sustainable Energy Action Plan** focuses on measures to reduce the city's impact on climate change. The campaign identifies what action is required locally to reduce the city's carbon emissions in order to meet UK targets. It has developed a programme to reduce Swansea's use of fossil fuels by 5% by increasing energy efficiency and developing renewable energy technology capacity.

Pressure groups and individuals

According to **Greenpeace** 150,000 people are dying every year because of climate change and, within 50 years, a third of all land-based species could face extinction. It seeks to spur national government and international organisations into action, with campaigns against the third runway at Heathrow and the development of new coal-fired power stations. Greenpeace supports energy conservation and renewable energy options to mitigate climate change.

Friends of the Earth was set up in 1971 and campaigns for a safe climate locally, nationally and internationally. It seeks solutions to environmental problems based on credible research and uses its network to involve over 1 million people to work for a more sustainable future.

The **Global Climate Campaign** is the collective name for all the organisations, groups and individuals around the world that come together for the **Global Day of Action**. This has taken place each year since 2005 at the same time as the annual **United Nations Talks on Climate Change**. People from all over the world join together in the world's major cities in over 100 countries to demand urgent action on climate and climate justice from governments meeting at the talks.

Climate Camp is a fast-growing, grass-roots movement of ordinary people taking action on climate change. Its members are impatient with government rhetoric and corporate spin, which they consider is holding up action needed to combat the problem. The camp was founded in August 2006 when 600 people gathered to protest at the UK's biggest single source of carbon dioxide, the Drax coal-fired power station in West Yorkshire.

Stop Climate Chaos Coalition is the UK's largest group of people dedicated to action on climate change and limiting its impact on the world's poorest communities. Its supporter base of over 11 million people combines more than 100 organisations, from environment and development charities to unions and faith, community and women's groups. The coalition is demanding practical action in the UK to prevent global warming rising beyond the 2°C threshold.

The politician Nigel Lawson set up a climate-change sceptic think tank in November 2009 called the **Global Warming Policy Foundation**. Its main purpose is to bring reason, integrity and balance to a debate that it believes has become unbalanced, alarmist and often intolerant. The group seeks to restore balance and trust in the climate debate, which it considers to be distorted by prejudice and exaggeration.

Crucially, climate-change sceptics are often funded by oil companies. Powerful business interests, especially in the USA, do not want to adopt measures that might inhibit their activities.

> **Examiner tip**
> Go to **www.thegwpf. org/** to see the arguments put forward by the sceptics. A balanced examination response will consider sceptical viewpoints as well as those believing we need to take drastic action immediately.

> **Knowledge check 14**
> What is the Green Deal?

1.6 How successful have strategies been in tackling climate change?

Evaluation of attempts to reduce climate change

A wide range of strategies can be employed to reduce climate change. Although these measures may have some effect, the steady rise in carbon emissions from developing nations is limiting their success. Methods to reduce emissions include:

- quotas on fossil fuel production such as those agreed by OPEC
- switching to alternative energy sources such as nuclear power or a mix of renewable options such as HEP, wind turbines, wave and tidal energy and solar panels
- improving energy efficiency and conservation in urban planning, building design and transportation
- developing carbon capture and storage technology
- reducing methane from waste materials
- afforestation and reforestation

More futuristic and ambitious schemes have also been suggested to limit climate change, such as seeding the oceans with iron to increase the uptake of carbon dioxide and building giant reflectors in space to reduce incoming solar radiation.

Industrialised countries contain about a quarter of the world's population, but they produce two-thirds of the world's energy-related carbon dioxide emissions. However, developing nations are likely to suffer the worst impacts of global warming. Governments and businesses that do not implement appropriate climate protection measures ought to bear some responsibility for the consequences of climate-induced damage by either reducing their emissions or paying into a compensation fund.

There has been a series of international meetings and agreements to deal with climate change and its consequences, but real progress is slow as each country seeks to protect its own interests. Multilateral agreements have to ensure sufficient levels of participation, get adequate action from all parties and encourage (or enforce) compliance in order to succeed.

The **Kyoto Protocol** has failed to live up to expectations despite the fact that, by November 2009, 187 states had signed up to it. This is because:
- the USA and Australia did not sign and ratify the treaty
- countries have not met their agreed aims to reduce their greenhouse gas emissions by an average of 5.2% from 1990 levels
- many countries, such as Russia, were not required to make reductions

Kyoto seems to have failed as carbon dioxide levels in the atmosphere are rising at a faster rate with no sign of slowing. Some argue that the protocol did not go far enough to curb greenhouse gas emissions and that the limits did not include emissions by international aviation and shipping.

Without the USA ratifying the protocol or rapidly developing economies such as China reducing emissions drastically, the targets cannot be met. Developing nations such as Brazil, China and India have no obligation other than monitoring and reporting emissions. However, Kyoto did establish a framework for future climate agreements and allowed countries to experiment with capping and trading emissions for the first time. It can therefore be seen as a limited and cautious first step, rather than a complete failure.

Examiner tip

Since this book was written the UN Climate Change Conference was held in Durban, ending in December 2011. Research the outcomes of this conference: did it make progress?

At the **UN Climate Change Conference in Bali** in December 2007 more than 10,000 participants, including representatives of over 180 countries, met with observers from intergovernmental and non-governmental organisations (NGOs) and the media. They agreed to the Bali Road Map for a new negotiating process that will lead to a post-2012 international agreement on climate change (when the Kyoto Protocol expires). There were decisions on technology transfer and reducing emissions from deforestation. An Adaptation Fund was also agreed to improve the defences of poor countries that lack the money, technology and human resources to cope with climate change. Funding comes from a 2% levy on revenues generated by the carbon credits scheme.

The **UN Climate Change Conference in Copenhagen** in December 2009 aimed to achieve accord between 77 countries to cut greenhouse gases. However, talks broke down and a watered-down agreement, the Copenhagen Accord, was signed by just five nations.

Most recently, climate scientists have been called to account following a number of controversies:

- 'Climategate', which involved emails of the University of East Anglia's Climate Research Unit being hacked into. Climate-change sceptics alleged that the emails showed evidence that data had been manipulated to prove that people were causing climate change.
- 'Glaciergate' revealed that the 2007 IPCC report had mistakenly drawn evidence from a WWF campaign suggesting that the Himalayan glaciers would disappear by 2035. This information was not based on peer-reviewed scientific assessment and proved to be inaccurate.

Examiner tip

Questions that ask you to evaluate or assess strategies expect you to look at a range of points and justify those that you think have the greatest impact.

Summary

- Understanding the global pattern of climates and their related biomes is a necessary prerequisite for understanding climate change.
- There are both long- and short-term climate changes. The reasons for the changes differ and have different causes.
- There is a growing body of evidence for current climatic change. The processes of change are related to environmental and human factors.
- The effects of climate change are debatable, although there is evidence from shifting climate belts, rising sea levels, retreating glaciers and extreme weather, all of which impact on societies. The impacts vary according to location and levels of development.
- International organisations, governments at various levels and ordinary people are developing strategies to address climate change.
- Despite the degree of uncertainty about the future effects of climate change, society needs to evaluate the success or otherwise of attempts to reduce the impact of climate change.

Theme 2: Investigating tectonic and hydrological change

2.1 What are the processes associated with plate tectonics?

Patterns of plates and plate boundaries

In 1912 Alfred Wegener published his theory that the continents were moving across the Earth's surface. He suggested that, 300 million years ago, they had formed a single vast continent that he named **Pangaea**. His theory of 'continental drift' was supported by geological, climatological and biological evidence, including:

- the continents of Africa and South America seem to fit together
- similar 290-million-year-old glacial deposits can be found in South America, Antarctica and India
- similar geological sequences are found in Scotland and eastern Canada

- similar fossils of reptiles, invertebrates and plants are found on different continents
- coal reserves are found in Antarctica so the climate there must once have been hot

Wegener believed that Pangaea split into two continents, Laurasia and Gondwanaland, about 200 million years ago and then continued to break up into their present-day arrangement.

Examiner tip

Make a list of the evidence for plate movement under the headings: geological, climatological, and biological. Research further details of the evidence. This will help you to learn this topic well.

As scientific technology improved during the twentieth century, more evidence was found to support the **theory of plate tectonics**. It was discovered that the sea floor was spreading in the mid-Atlantic. Reverses in the Earth's polarity are recorded in the new crust, creating a striped pattern that is mirrored on either side of the mid-Atlantic ridge. To compensate for this growth in the Earth's crust at constructive plate margins, a tectonic plate elsewhere must be being destroyed. Huge ocean trenches were discovered where the ocean floor was being subducted below the lighter, less dense continental crust.

There are seven major tectonic plates — African, Antarctic, Eurasian, Indo-Australian, North American, South American and Pacific — and several smaller ones such as the Nazca and Philippines plates (Figure 19). Oceanic plates are 50–100 km thick and the rock forming them is no older that 180 million years. Continental plates are 100–250 km thick and, in places, much older. Rates of movement vary at between approximately 0.60 cm per year to 10 cm per year.

The Earth's crust and the outer part of the mantle, consisting of more rigid material, form the **lithosphere**. The lower part of the mantle, which is molten or semi-molten, is called the **aesthenosphere**.

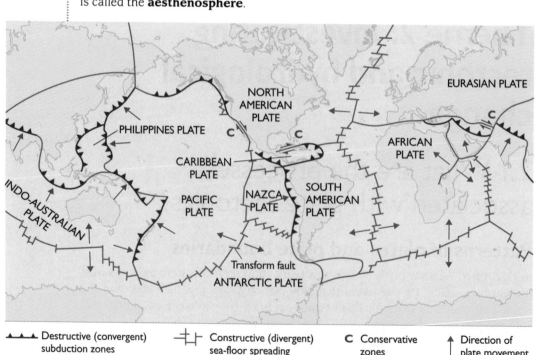

Figure 19 The global pattern of tectonic plates

Processes associated with constructive, destructive and conservative plate margins

The movement of tectonic plates is driven by thermal convection currents in the upper mantle using heat derived from the radioactive decay of minerals deep within the Earth and residual heat from the Earth's formation. This heat causes plumes of hot magma to rise. If the crust is thinner on the mid-ocean floor, the hotter and less dense magma breaks through to form new crust as it is cooled by the water.

The actual driving mechanism of plate movement has not been determined. Until the early 1990s scientists thought that mantle convection, sea-floor spreading and magma intrusion at mid-ocean ridges drove plate movement in a process called **ridge push**. However, more recently the concept of **slab pull** — in which a dense ocean plate sinks into a subduction zone because of the pull of gravity — has gained precedence. The plates are hottest at the mid-ocean ridges and cool, becoming more dense and heavier, as they move away.

Knowledge check 15

Name two major and two smaller tectonic plates.

Constructive (divergent) margins

At constructive plate boundaries (Figure 20), the plates move away from one another. Magma rises to the surface, forming oceanic crust at oceanic ridges or continental crust at rift valleys. This type of plate margin is characterised by shallow-focus earthquakes and volcanic activity. Iceland lies across the **mid-Atlantic ridge** and new islands, such as Surtsey in 1963 and Heimey in 1973, have been formed due to volcanism. In 2010 Eyjafjallajökull erupted along this plate boundary.

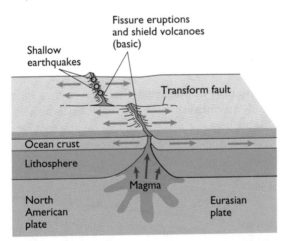

Figure 20 Constructive margin processes

Destructive (convergent) and collision margins

There are three types of convergent boundary at which plates move towards each other: oceanic–continental, oceanic–oceanic and continental–continental. The first two are destructive plate boundaries where one plate is subducted beneath the other. Collision margins are where two plates both consisting of continental crust move together.

At **oceanic–continental boundaries**, denser oceanic crust is subducted beneath lighter continental crust (Figure 21), creating deep **ocean trenches** and volcanic **mountain chains** and **batholiths** beneath the surface. The ocean plate is partially melted, producing andesitic or granitic magma. Shallow-to-intermediate-focus earthquakes commonly occur at these margins such as that in Chile in February 2010, which was caused by the subduction of the Nazca plate beneath the South American plate.

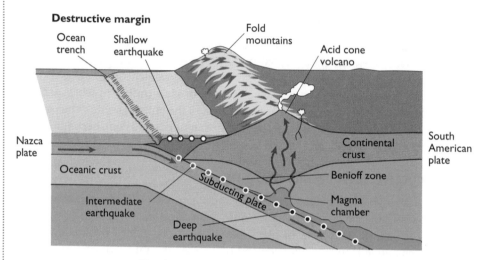

Figure 21 Destructive margin processes

At **oceanic–oceanic boundaries**, one plate is subducted and a volcanic island arc with an adjacent deep-sea trench may form and shallow-to-deep-focus earthquakes occur from the trench towards the **island arc**, in an area known as the **Benioff zone**. The Aleutian Islands form an arc from Alaska in the USA to the Kamchatka Peninsula in Russia along the northern edge of the **Pacific Ring of Fire**.

At **continental–continental boundaries**, the crust cannot subduct so it is forced upwards to form **fold mountains** such as the Himalayas (Figure 22). Shallow-focus earthquakes occur here but there is almost no volcanic activity. The Kashmir earthquake in Pakistan in 2005 occurred in the region of the colliding Eurasian and Indo-Australian plates.

Examiner tip

You should know the names of major plates and the direction in which they are travelling. Valuable marks are lost by students who confuse their plate boundaries. Practise sketching annotated diagrams of the three types of plate margin so you can add them to your exam answers.

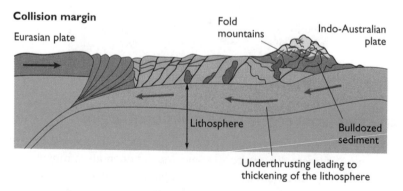

Figure 22 Collision margin processes

Conservative (transform) margins

At conservative boundaries the plates slide laterally past one another (Figure 23). There is no volcanic activity but shallow-focus earthquakes are common along faults running parallel to the plate margin. The San Andreas Fault, along the west coast of North America where the Pacific and North American plates meet, puts a string of major cities such as San Francisco and Los Angeles at risk.

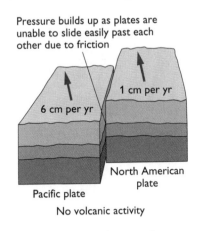

Pressure builds up as plates are unable to slide easily past each other due to friction

1 cm per yr

6 cm per yr

North American plate

Pacific plate

No volcanic activity

Figure 23 Conservative margin processes

Knowledge check 16

Distinguish between the three types of plate boundary.

2.2 What are the hazards associated with tectonic events?

Earthquakes and volcanoes are the primary hazards caused by tectonic activity. Secondary impacts include tsunamis, mudslides and short-term climate change. Tectonic hazards have local and regional impacts that can have demographic, economic and social consequences.

Volcanoes and their impacts

Volcanoes occur at constructive and destructive plate margins, or over **hotspots** in the Earth's crust, where hot magma, ash and gases escape from below the Earth's surface. The type of eruption and the nature of its products depend on the location of the volcano.

Volcanoes can be classified by their size and shape or by the nature of their eruptions. However, these features are connected as the type of eruption affects the consequent landform. Eruptions may be categorised as: **effusive eruptions**, which involve the outpouring of magma that is relatively low in viscosity and in gas content, and **explosive eruptions**, which involve magma that is more viscous and acidic.

There are two basic types of volcano. **Shield volcanoes** (Figure 24), such as Mauna Loa in Hawaii, have a flatter shape and wider diameter as layers of liquid lava spread out over a wide area. They produce lavas that are hotter and less viscous. Gas release is rare, so the gentle slopes are built up with layers of basaltic lava. Often the crater remains open and filled with bubbling lava and typically there is just one central vent. The Hawaiian volcanoes are formed as the result of the islands' location over a hotspot in the centre of the Pacific plate. Shield volcanoes may also be found at constructive (divergent) plate margins, forming rift volcanoes with open fissures, as in Iceland. The speed of the lava flow and, in Iceland, the flooding caused as ice is melted, are the main hazards.

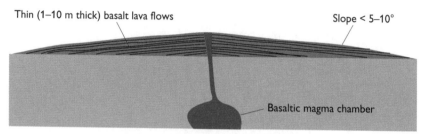

Thin (1–10 m thick) basalt lava flows

Slope < 5–10°

Basaltic magma chamber

Figure 24 Cross-section of a shield volcano

Stratovolcanoes or **composite volcanoes** (Figure 25), such as Mayon in the Philippines, are more pyramidal in cross-section and are built up with layers of lava, rock fragments and ash. They produce lavas that are cooler and more viscous, allowing gas pressure to build up between infrequent eruptions. This causes explosive eruptions. These volcanoes contribute to the formation of volcanic arcs at destructive plate margins and are sometimes called subduction volcanoes. Examples of stratovolcanoes include Mount Fuji in Japan, Merapi in Indonesia, Galeras in Columbia and Cotopaxi in Equador, all located around the Pacific Ring of Fire. The sequence of slow-flowing lava and pyroclastic material erupted produces steep layers of ash and lava domes. This increases the chance of landslides and avalanches. A common hazard at stratovolcanoes are **lahars** or mudflows, which are dangerous because they move quickly and are usually hot. A **caldera** may be formed in the greatest eruptions, when the top of the symmetrical cone is blown off leaving a collapsed central crater.

Examiner tip
Draw annotated sketch diagrams of the different types of volcano and include the names of located examples of these volcanoes. This will help you to memorise the details in preparation for the exam.

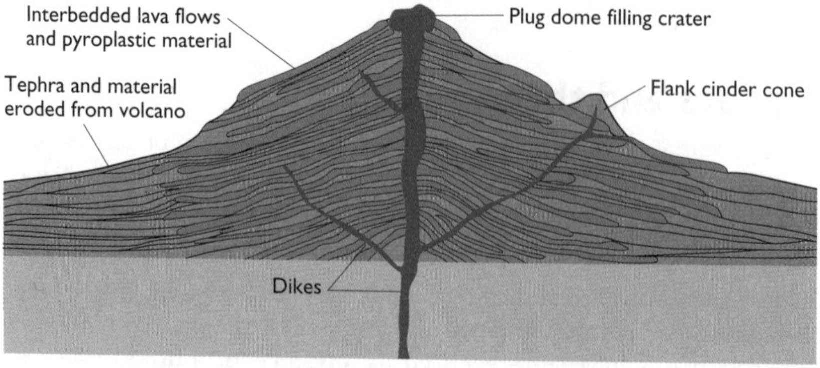

Interbedded lava flows and pyroplastic material

Plug dome filling crater

Tephra and material eroded from volcano

Flank cinder cone

Dikes

Figure 25 Cross-section of a stratovolcano

Volcanoes vary according to how explosive they are and the volume of their eruptions. The main types of volcanic event are summarised in Figure 26. Hawaiian volcanoes such as Kilauea do little damage as they are rarely explosive. Krakatoan eruptions are cataclysmic explosions throwing volcanic dust tens of kilometres into the atmosphere and sometimes causing tsunami.

The more explosive volcanoes often create **pyroclastic flows**. These are dangerous, fast-moving flows of hot gas and rock. Temperatures may reach 1,000°C and the velocity of the flow may be up to 725 kilometres per hour. Recent examples of such devastating eruptions include Mount St Helens in 1980 and Mount Pinatubo in 1991.

Knowledge check 17

What is pyroclastic material?

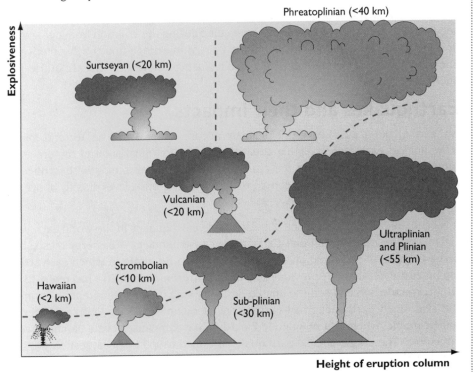

Figure 26 A classification of types of eruption

Volcanic events

The worst volcanic eruptions in modern times, in terms of people killed, were as follows:

- Tambora, Indonesia (1815) — 60,000
- Krakatau, Indonesia (1883) — 36,600
- Mount Pelée, Martinique (1902) — 29,000
- Nevado del Ruiz, Columbia (1985) — 23,000

Each of these eruptions caused pyroclastic flows — one of the deadliest hazards associated with volcanic activity — and additional hazards increased the numbers of casualties. The Mount Pelée eruption generated massive pyroclastic flows that destroyed the coastal town of Saint Pierre. Almost everyone who lived there was burned alive, suffocated or buried by the fast-moving ash flow.

When Nevado del Ruiz erupted in 1985, 23,000 people were killed in the lahar that followed. The eruption and pyroclastic flows were quite small but they melted much of the ice cap at the summit of the volcano, causing a surge of water, sediment, rock and vegetation that deposited up to 5 m of mud on the town of Armero below.

The 1883 eruption of Krakatau was heard almost 2,000 km away and produced pyroclastic flows and ash falls that killed about 4,600 people. However, the worst hazard was the **tsunami** that followed the collapse of the flanks of the volcano into the ocean. Without warning 32,000 people were killed in the low-lying coastal communities surrounding the island volcano.

The eruption of Tambora in 1815 — the most powerful in recorded history — ejected enough volcanic material into the atmosphere to cool the atmosphere. About 11,000 people living near the volcano in Indonesia died during the eruption from pyroclastic flows and ash falls. The following year is known in Europe and North America as 'the year without a summer' as crops failed causing famine and disease and killing another 49,000 people.

Earthquakes and their impacts

An earthquake results from the sudden release of pressure as tectonic plates move against each other, creating **seismic waves**. The initial point of rupture is the **focus** or **hypocentre** within the crust, so that earthquakes may be described as deep or shallow focus. The **epicentre** is the point at ground level directly above the focus.

It is estimated that there are 500,000 detectable earthquakes in the world each year: 100,000 of these can be felt and 100 of them cause damage. The energy released by an earthquake is measured on a **moment magnitude scale** in which each step is ten times greater than the previous step on a logarithmic scale. This has replaced the **Richter scale**, which used the amplitude of shaking on a seismograph to estimate the size of an earthquake. An earthquake of magnitude 3 or lower is generally imperceptible, whereas a magnitude 7 earthquake can cause severe damage over a large area. The magnitude 9.5 earthquake in Chile in 1960 is the largest earthquake ever to be measured.

The intensity of shaking is measured on the **Mercalli scale**. This is a descriptive means of judging the extent of damage that relies on people's interpretation of the events so, if no-one is there, nothing will be measured. Different underlying rock and sediments will give varying results for the same magnitude. This makes it difficult to compare earthquakes with this scale, which runs from I ('instrumental', i.e. detected only by seismographs) to XII ('catastrophic').

The maximum shaking in the 2010 Chile earthquake was around 35 times greater than that of the Haiti earthquake a few weeks earlier. About 500 times more energy was released in the Chile earthquake, it lasted longer and affected an area some ten times larger. However, as Table 3 shows, the human impacts did not differ to the same extent. The closer the focus of the earthquake is to the surface, the greater the shaking and potential for damage to surface structures.

If a large earthquake's focus is located beneath the seabed and it causes displacement, this may cause a **tsunami**. On 26 December 2004 an undersea earthquake off the west coast of Indonesia and with a magnitude of 9.3 (the second largest ever recorded) triggered a series of tsunamis affecting coastlines surrounding the Indian Ocean, killing nearly 230,000 people.

Earthquakes can set off **landslides** and, occasionally, volcanic activity. In 1970 a large earthquake beneath the Pacific Ocean off the coast of Peru loosened rocks and ice on Mount Huascarán, 130 km away. This caused a huge landslide that reached speeds of over 200 km per hour. It badly damaged the town of Ranrahirca 12 km from the mountain and destroyed the village of Yungay; 67,000 people died and 800,000 lost their homes.

Table 3 A comparison of the Haiti and Chile earthquakes

	Haiti	Chile
When	Tuesday 12 January 2010 at 4.53 p.m.	Saturday 27 February 2010 at 3:34 a.m.
Magnitude	Magnitude 7.0. The earthquake shook a small area very intensely. The area included Port-au-Prince with 2 million inhabitants and buildings that were not built to withstand earthquakes.	Magnitude 8.8, the fifth strongest ever recorded. Described as VIII (destructive) on the Mercalli scale. The earthquake was felt 2,900 km to the east in São Paulo, Brazil.
Epicentre	16 km west of Port-au-Prince at a depth of 10 km — the most powerful to hit Haiti for 200 years.	90 km northeast of Concepción at a depth of 35 km.
Plate boundaries and movement	The earthquake occurred along a strike-slip fault (the same type as California's San Andreas Fault) marking the boundary between the North American and Caribbean plates (Figure 19). The Caribbean plate moves east at about 2 cm a year but pressure had built up as the fault had not moved significantly for 200 years.	The earthquake took place between the Nazca and South American plates (Figure 19), where they converge at a rate of 80 mm a year. The segment of the fault that ruptured was some 640 km long, lying just north of the 1,000 km segment that ruptured in the great earthquake of 1960.
Secondary impacts	33 aftershocks ranging in magnitude from 4.2 to 5.9. The tsunami had recorded wave heights (peak-to-trough) of 12 cm at Santo Domingo in the Dominican Republic and 2 cm at Christiansted in the Virgin Islands.	More than 100 aftershocks, with several above magnitude 6.0. The tsunami was recorded with amplitudes of up to 2.6 m high at Valparaíso in Chile, 48 cm in Hawaii and 15 cm in Alaska. Numerous landslides occurred in the coastal mountains, there was liquefaction in saturated sand dunes, and extensive ground cracks and subsidence throughout most of the area. The Earth moved 8 cm off its axis.
Casualties	230,000 people killed, 300,000 injured, 1.1 million displaced.	521 people killed, over 1 million displaced.
Estimated cost	$8 to $14 billion reconstruction costs. About 20% of the structures in the city of Port-au-Prince collapsed and up to 80% of those still standing suffered serious damage.	$4 to $7 billion; more than 0.5 million houses were damaged.
Buildings destroyed or damaged	280,000. Haiti has no real construction standards.	500,000. A strict 'seismic design' building code has existed since 1960, but a much larger area was affected.
Socioeconomic data	• Population: 9 million. • HDI: 0.532 (149th). • 80% below the poverty line. • Life expectancy: 61. • GDP per capita: $1,300. • Most Haitians live on less than $2 a day and over two-thirds do not have formal jobs.	• Population: 16 million. • HDI: 0.878 (44th). • 18% below the poverty line. • Life expectancy: 78. • GDP per capita: $14,700.

Large earthquakes can trigger **volcanic eruptions** up to 500 km away as well as many months later. After the 1960 Chile earthquake there were six eruptions in the following 18 months — a significant increase on the average eruption rate of about one per year. The Kilauea volcano in Hawaii erupted less than an hour after an earthquake hit a few kilometres away in 1975. In 2006 a magnitude 6.4 earthquake hit Java in Indonesia. Three days later geologists studying Merapi and **Semeru** volcanoes, located 50 and 280 km away, recorded magma flowing inside both volcanoes at twice the normal temperature and speed.

2.3 How are tectonic hazards perceived and managed?

Different perceptions and awareness of tectonic hazards by groups with conflicting interests

People choose to live in potentially hazardous places because:
- hazards are unpredictable
- the level of risk may have increased since they moved there
- they accept hazards as an inevitable possibility
- the benefits of living there outweigh the hazard costs
- they are unable to move elsewhere for social, economic, cultural or political reasons

People's perception of risk varies according to a variety of factors such as their level of education, culture or ethnicity, socioeconomic status and past experience of hazardous events. These can affect their understanding, preparedness and actions in the immediate aftermath of a hazardous event. In the longer term, their perception of the level of risk will determine their response. Residents of more affluent countries are usually better able to cope with the consequences of tectonic hazards. In societies where there is a disparity of wealth, there may be conflict between different groups as access to safety and security may vary.

Engineers and scientists calculate risk as the degree of hazard multiplied by the cost, whereas social scientists consider risk in terms of the vulnerability or resilience of people and communities to the hazard. Vulnerability is the tendency for something or someone to be damaged and resilience refers to their ability to resist or recover from damage.

The risk equation measures the level of hazard for an area:

$$\text{risk (R)} = \frac{\text{frequency or magnitude of hazard (H)} \times \text{level of vulnerability (V)}}{\text{capacity of population to cope and adapt (C)}}$$

To what extent hazards are feared or accepted as a natural part of life (known as **fatalism**) depends on people's understanding of the level of risk; this will be affected by local culture and experience whereas level of education may determine how people respond to a hazard. Those who are fearful may move away to a place they perceive to be safer, perhaps temporarily; those who are fatalistic will stay and accept whatever happens. Those living in societies with both technical know-how and the money to invest in preparation for and protection against hazardous events will be less fearful of or fatalistic about the outcomes.

Strategies to manage tectonic hazards and their effectiveness

The management of hazards is partly influenced by the perception of risk. There was no warning system in place in the countries around the Indian Ocean at the time of the 2004 tsunami because such events were perceived as less common in that region. A system was installed 18 months later after over 220,000 people lost their lives.

In Chile in 2010, villagers complained that the tsunami warning was inadequate. The installation of such systems is determined by cost–benefit analysis so that poorer countries take the risk that a significantly damaging event will not occur. Tsunamis are far more common in the Pacific Ocean; Japanese citizens, for example, receive frequent warnings. Levels of preparedness and response are closely related to the affluence of a country and to its past experience. This was clearly shown in the differing impacts of the earthquakes in Haiti and Chile (Table 3, page 43).

There are three stages in the management of tectonic hazards: preparation, immediate response and longer-term response.

Preparation

- Store emergency equipment such as tents, clean water, food supplies, first aid and search-and-rescue gear.
- Train specialist workers for a hazardous event.
- Raise public awareness by having earthquake drills, for example.
- Set up warning systems and, if possible, evacuation plans.
- Set up insurance to provide finance for reconstruction after the event.
- Ensure households have emergency kits ready with water, food, clothing, a torch, a radio, spare batteries, first aid, etc.

Immediate response

- Coordinate search-and-rescue.
- Establish medical reception centres.
- Establish alternative accommodation and emergency shelters.
- Provide clean water, wood and sanitary facilities.
- Set up communication channels.
- Ensure householders and businesses turn off gas supplies to prevent fire.

Longer-term response

- Construct safer buildings such as earthquake-resistant structures using flexible steel frames, shock absorbers, cross-bracings and counterweights.
- Improve prediction and monitoring systems (for volcanic activity and the secondary hazards of earthquakes).
- Carry out land use zoning — use the areas that are most at risk for recreation, open spaces or low-rise buildings.
- Provide aid and government support for reconstruction.

Figure 27 models the impact of a disaster from pre- to post-disaster. It also considers the role of emergency relief and rehabilitation.

Examiner tip
Research the 11 March 2011 tsunami that followed the level 9.0 earthquake off the coast of Japan. What were Japanese perceptions of the hazard and were the strategies sufficient in Japan and elsewhere? This was the fifth-strongest recorded earthquake and the examiner will expect you to know about it.

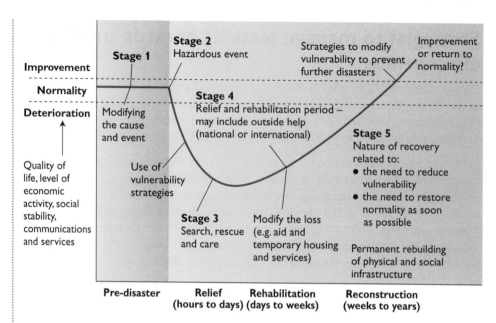

Figure 27 A disaster-response curve

Responding to a tectonic event

On 11 March 2011, an earthquake took place at 14:46 JST some 67 km from the nearest point on Japan's coastline; the tsunami took at least 10 minutes to reach land. At 15:55 JST, the tsunami flooded Sendai Airport. The Earthquake Early Warning System sent out warnings now believed to have saved many lives and the Japan Meteorological Agency warned of a major tsunami at least 3 metres high. The actual heights were, in places, nearly 10 metres and inundated an area of approximately 561 km².

In the immediate aftermath, the Japanese government mobilised the Self-Defence Forces, and many countries sent search and rescue teams. Over 300,000 people were displaced by the tsunami in the Tohoku region, and there were shortages of food, water, shelter, medicine and fuel for survivors. Over $1 billion of aid was donated worldwide. Initially, large shelters were set up; these were replaced over time with smaller, temporary units. A strong police force ensured public disorder was minimised, though there was some looting.

In the longer term, the economic impact, including loss of industrial production and the cost of rebuilding, has been estimated at $122 billion. The most striking impact was the damage to the Fukushima nuclear plant and the reduction in supply to the national grid; residents were evacuated within a 20-mile radius of the plant. Factories closing on different days of the week, rolling blackouts and switching off unnecessary street lighting reduced electricity use. A massive clear up began immediately to remove the debris of smashed buildings and vehicles left behind the tsunami.

Although rich countries such as the USA and Japan can usually cope with the consequences of disasters using their own resources, poorer nations are reliant on the support of international organisations, charities and other countries to help out

Examiner tip

Find out how Japan has recovered since the 2011 earthquake. What have been the global consequences?

in the immediate aftermath of a hazard event and the longer-term reconstruction. The **Disasters Emergency Committee** (DEC) coordinates 13 humanitarian aid agencies, bringing together aid, corporate, public and broadcasting sectors to raise funds to deliver effective and timely relief. One such organisation is the British Red Cross, which responds to natural disasters alongside the International Red Cross.

The **United Nations International Strategy for Disaster Reduction** (UNISDR) aims to build disaster-resilient communities by promoting increased awareness of the importance of disaster reduction to limit human, social, economic and environmental losses due to natural hazards. Often it plays a key role in coordinating rescue and support services as people and organisations from around the world, such as the British Fire Brigade and Médecins Sans Frontières (MSF), arrive to help. However, it is recognised that reducing vulnerability to hazards is more cost-effective than large-scale clean-up operations. A key role of the UNISDR is to improve the preparedness for and resistance to the consequences of natural disasters.

Integrated risk management identifies hazards, analyses the risks, sets priorities and implements risk reduction plans. Strategies can be subdivided into prediction, prevention and protection.

Prediction of tectonic hazards is limited to monitoring eruptions when they show signs of occurring and warning about the secondary consequences of earthquakes such as tsunamis. Earthquake-prone areas are known, but earthquakes cannot yet be predicted.

Prevention is not an option for volcanoes or earthquakes other than to evacuate people from the area affected in the case of impending volcanic eruptions.

Protection is possible as sometimes the event can be managed to limit damage. Successful schemes include digging trenches and building artificial barriers to limit the impact of lava flows and lahars, using dynamite to divert flows of lava on the slopes of Mount Etna in Italy, and pouring sea water on lava to solidify it in Iceland. These are smaller-scale but reasonably successful actions.

The 2010 volcanic ash crisis following the eruption of Eyjafjallajökull in Iceland was probably the first time that many Europeans became aware of the impact of volcanoes on modern societies. In Iceland the fear was of a Jokalaup, the melting of the snow and ice resulting in flash floods. However, this did not occur. Figure 28 shows the spread of the ash cloud on 10 May and explains why on that day flights to Spain and Portugal were disrupted and transatlantic flights were diverted, adding up to 2 hours to flight times. In this case protection involved stopping flights for 6 days in mid-April 2010. This had high costs, including:

- loss of revenue to airlines, which had to refund fares and pay for longer stays at destinations
- loss of revenue to airports from car parking, landing fees and profits from retail rentals
- loss of revenue to tax authorities (departure tax)
- loss of revenue to retailers in airport terminals
- the effects of people being away from work because they were stuck abroad
- insurance claims and payouts

> **Examiner tip**
> Look up the Montserrat Volcano Observatory at **www.mvo.ms/** and study how it is monitoring the Soufriere Hills Volcano in the region and its social impact. This research will help you to understand how we live with impending disasters.

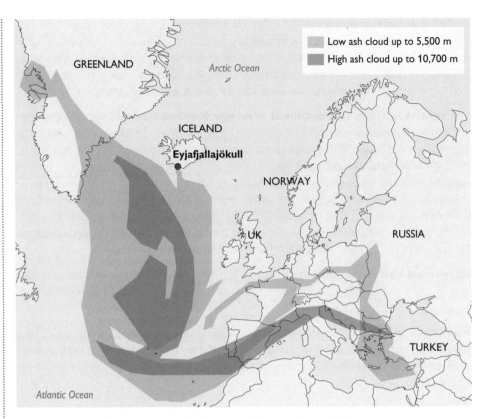

Low ash cloud up to 5,500 m
High ash cloud up to 10,700 m

Figure 28 The distribution of the ash cloud, 10 May 2010

Knowledge check 18

What is a Jokalaup?

Summary

- Plate tectonics involves a set of processes associated with different types of plate boundary.
- There are hazards associated with tectonic activity, which include tsunamis, flooding and landslides. There are also demographic, economic and social effects. All effects can be studied both at the local and regional scale. The reach of some events can be global.
- Tectonic hazards are perceived differently by those affected and those beyond the affected area. Perceptions often guide policy, which has to take into account the differing interests of groups.
- Strategies are developed at local, regional, national and international levels to mitigate the effects of tectonic hazards.

2.4 What are the hydrological processes associated with drainage basins?

The drainage basin system: inputs, flows, stores and outputs

A drainage basin is an area of land that is drained by a river and its tributaries. Water and sediment are transferred downslope from high land to an ocean, sea or lake. The drainage basin system has measurable inputs, flows, stores and outputs, and it is that part of the hydrological cycle that involves the land (Figure 29). It is an open system as there are inputs from and outputs to the atmosphere and to the sea, whereas the hydrological cycle is a closed system.

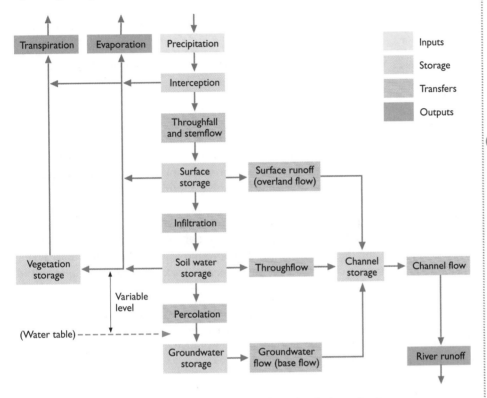

Figure 29 The hydrological cycle in the drainage basin

Examiner tip
Figure 29 is a key diagram and you should memorise it to ensure you use the terminology correctly in the exam.

Inputs

The input to the drainage basin system is **precipitation**. This is water and ice that falls from the clouds in the form of rain, snow or hail.

Flows

The flows (transfers) between stores in the drainage basin system are as follows:
- **Throughfall** and **stemflow** and **drip** occur through the vegetation (for example, running down the trunk and as trees dripping from leaves after a rain storm).
- **Infiltration** is the movement of water from above to below the surface of the ground. The rate of infiltration depends on the amount of water already in the soil, the porosity and structure of soil, and the type and extent of vegetation. After a long, dry period the ground surface may be 'baked' and relatively impermeable.
- **Percolation** is the movement of water down through the soil and permeable rock.

Water also flows towards stream and river channels from surface, soil and groundwater stores:

Knowledge check 19

What is the difference between an open and a closed system?

- **Surface runoff** or **overland flow** travels across saturated or impermeable land surfaces. Surface runoff or overland flow depends on the ability of ground surface to absorb water.
- **Throughflow** is the movement of water through the soil towards the surface, often emerging on valley sides to form springs.
- **Groundwater flow** or **base flow** is the movement of water through rocks towards the surface.

In hot climates, water may be drawn upwards through the soil by **capillary action**.

Stores

In sequence, stores include the following:

- **Interception** as vegetation and other surfaces above the ground catch falling precipitation. Interception by vegetation may remove up to 30% of water from the system. Water may be evaporated from here or continue falling to the ground.
- **Surface storage**, which includes any body of water from a small puddle to a large lake.
- **Soil water storage**, which is essential for plant growth.
- **Groundwater storage** in permeable rocks. Large stores of groundwater are called aquifers and the surface of this underground water is the water table. Water stored in the ground collects above impermeable rock in porous rock and soil to create a zone of saturation.

Examiner tip

You need to be able to define all these terms, as they are essential to understanding the hydrological cycle.

The **river channel** is also a store, combining the water that flows into it from the surface, soil and ground. The water then flows in the channel to become **river runoff** or discharge. About 5% of total precipitation can be attributed to 'channel catch' — rain falling directly into the river.

Outputs

The outputs are:

- **river runoff** as water is carried overland to the sea
- **evaporation** as water is changed to water vapour
- **transpiration**, which is the evaporation of moisture through the stomata on leaf surfaces

Evaporation and transpiration are difficult to measure individually, so they are often combined as **evapotranspiration**.

Water budget

The balance between the water inputs and outputs is known as a water budget or water balance. As river runoff and precipitation are easily measurable, it is possible to estimate evapotranspiration and changes in storage using the formula:

$$Q = P - E \pm \Delta S$$

where Q is river runoff, P is precipitation, E is evapotranspiration and ΔS is change in storage.

The water budget changes throughout the year, as Figure 30 shows for an area in south Wales.

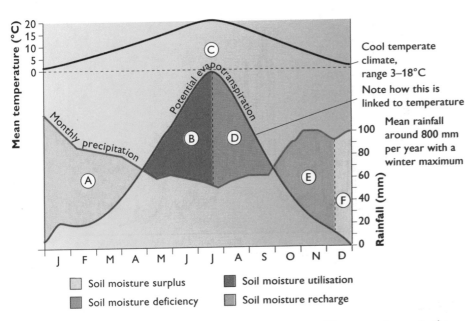

Cool temperate climate, range 3–18°C

Note how this is linked to temperature

Mean rainfall around 800 mm per year with a winter maximum

☐ Soil moisture surplus   ■ Soil moisture utilisation
■ Soil moisture deficiency   ☐ Soil moisture recharge

At **A** precipitation is greater than potential evapotranspiration. The store of water in the soil is full and therefore there is ample moisture for plant use, runoff into rivers and the recharging or refilling of the groundwater supply.

At **B** potential evapotranspiration is greater than precipitation and plants rely on stored groundwater. The store gradually diminishes.

At **C** the losses through evapotranspiration and take-up by plants have used up all the stored groundwater. If the area is being cropped, irrigation is necessary to maintain plant growth. When storms occur, the outcome will often be direct overland flow over the hard, baked ground. If infiltration is possible, water will percolate into the ground.

At **D** there is continuing soil moisture deficiency and plants have to adapt to survive. Adaptation can only take place over a long period of time. Irrigation is essential to maintain cropping.

At **E** precipitation is once again greater than potential evapotranspiration. The soil water store begins to refill again. Sometimes droughts continue and the store does not refill.

At **F** the soil water store is full and **field capacity** is reached. Surplus water will percolate into the groundwater store.

In other climatic regions the relationship between precipitation, temperature and potential evapotranspiration will alter the sequence above.

Figure 30 A water budget graph for a cool temperate area in south Wales

> **Knowledge check 20**
>
> State the difference between throughflow, percolation and infiltration.

The characteristics of river regimes and the physical and human factors influencing them

Variations in river runoff or discharge over the course of a year are called **river regimes**. These are illustrated by annual hydrographs showing the river discharge, usually in cubic metres per second or cumecs. Figure 31 shows the regime for the River Thames.

There are several factors that influence river regimes, including climate, geology, land use and river management.

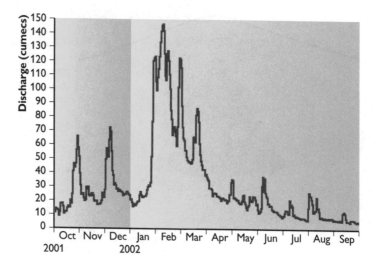

Figure 31 The annual river regime of the Thames at Reading, 2001–02

Climate

The volume of discharge in a river depends on the climate. Rivers in humid environments are fed by a constant flow of groundwater and are described as having **perennial flow** regimes. In drier climates, there may be insufficient water to maintain channel flow, so these rivers are described as **intermittent flow** rivers as they sometimes run dry. In the driest regions, channels may be fed by occasional flash flooding and are described as having **ephemeral flow** regimes.

If the regional climate has distinct wet and dry seasons, such as monsoon climates, this is reflected in the river regime, which may flood during the rainy season. The regimes of rivers fed by snow and glacier meltwater show distinct peaks and troughs in their annual hydrographs. Evapotranspiration is higher in summer than in winter if there is a significant seasonal difference in temperature throughout the year. This also affects the amount of water available for channel flow.

Geology

The rocks underlying a drainage basin can also influence its hydrology. **Permeable** rocks may be:

- porous, i.e. they contain pore spaces (e.g. sandstone and chalk)
- pervious, i.e. they contain joints and cracks through which water may travel (e.g. limestone)

Granite and basalt are **impermeable** so water can only pass through fractures in such rocks. River systems running over impermeable rocks will be more prone to flooding as there is less groundwater storage, whereas rivers running over permeable rocks will be more likely to dry up during droughts as water will be stored in the rock.

Land use

Human activity can change river regimes in a number of ways. The presence or absence of forestry determines the level of interception and so affects the time it takes for input from precipitation to work through the system and reach the river channel.

Regimes are also influenced by farming practices. Traditionally in the UK, farmers left their fields after harvest with the remains of the crop plants binding the soil. Fields were ploughed in late winter and sown with spring crops. The development of winter crops that are sown in late autumn means that fields are comparatively bare during the wetter winter months, reducing interception and increasing surface runoff. This leads to higher discharge and a greater risk of flooding.

Farmers use water for irrigation. In some parts of the world this has had a significant impact in the volume of water stored in aquifers and affecting the level of base flow.

In urban areas, rivers are more intensively managed to prevent flooding and make use of the water resources. The development of impermeable surfaces and storm drains may increase river discharge.

River management

Dam-building has the most significant impact on the regime of a river. Following the construction of a dam, the downstream flow of water in the river is controlled and rarely affected by seasonal variations in precipitation or meltwater. People also affect discharge by changing the channel: straightening meanders, deepening the channel and canalising it increases the velocity. Any form of river management has an impact on the hydrology of the river.

The characteristics of flood hydrographs and the human and physical factors influencing them

The volume of river discharge varies in different parts of the drainage basin because of a number of factors, which can be illustrated using a storm or flood hydrograph. Figure 32 illustrates variations in river discharge in response to a single rainfall event.

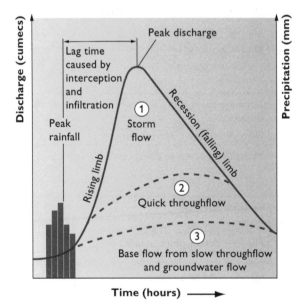

Key
1 Runoff from direct overland flow, or channel precipitation, reaches the stream very rapidly and leads to a rapid rise in discharge
2 Throughflow is slower, but nevertheless has an impact on the rising limb. It also contributes to the river when the storm flow has finished
3 Base flow may only arrive some 20–30 hours after the storm event, as the water has to seep into the system from deep below ground

Figure 32 A storm hydrograph

> **Knowledge check 21**
>
> What is the difference between perennial, intermittent and ephemeral flows?

> **Knowledge check 22**
>
> What is interception?

> **Examiner tip**
>
> You should be able to reproduce this diagram in an exam because it is the key to understanding flooding.

Examiner tip

Use the factors listed in Table 4 to describe a flood event you have studied.

Flood hydrographs show rainfall as a histogram and discharge as a line graph. The distance between **peak rainfall** and the **peak discharge** is called the **lag time**. It is a measure of the time it takes for the water to reach the river channel. This can be affected by underlying geology, vegetation and land use, as well as the shape and size of the drainage basin (Table 4).

Table 4 Comparing hydrographs

Factor	Short lag time, high peak, steep rising limb	Long lag time, low peak, gentle rising limb
Hydrograph shape	Described as **'flashy'**.	Described as **'flat'**.
Weather	• Intense storm with heavy rainfall, falling faster than infiltration rate.	• Steady rainfall, less than infiltration capacity of ground surface.
Climate	• Rapid snowmelt because of rapidly rising temperatures. • Low evaporation because of lower temperatures.	• Slow snowmelt because of lower temperatures or no snow. • High evaporation because of higher temperatures.
Geology	• Less permeable, granite, slate. • Little groundwater storage.	• More permeable, sandstone, chalk. • Greater groundwater storage.
Soil	• Less permeable clay soils. • Slow infiltration rate.	• More permeable sandy soils. • Fast infiltration rate.
Relief	• Steeper slopes causing rapid surface runoff and little infiltration and percolation.	• Gentler slopes and relatively flat, so slow or no surface runoff, more infiltration and percolation.
Basin geometry	• Smaller, circular. • Many smaller streams feeding river system.	• Larger, elongated. • Few smaller streams feeding river system.
Vegetation cover	• Bare or low density. • Deciduous forest in winter. • Deforested. • Little or no interception.	• Dense. • Deciduous forest in summer or evergreen forest. • High interception.
Antecedent (pre-existing) conditions	• Drainage basin is wet from previous rain, water table is high and soils saturated. • Little infiltration or percolation.	• Drainage basin is dry, water table is low and soils unsaturated. • High infiltration and percolation.
Land surface	• Wet (possibly with standing water) or dry (possibly baked hard after long, dry spell). • Reduced infiltration.	• Moist and porous enabling fast infiltration.
Land use	• Arable farming. • Urban. • Reduced interception and infiltration.	• Pastoral farming. • Rural. • Higher interception and infiltration.
Urbanisation	• High population density, large areas covered with tarmac and buildings. • Reduced infiltration.	• Low density population, mostly rural with little tarmac and few buildings. • High infiltration.

2.5 What are the causes and consequences of flooding?

Floods occur when the volume of rainfall or snowmelt exceeds the capacity of the drainage basin system and the excess water flows overland (Table 5). Flooding is a natural hazard that impacts on people and on the environment. Building on floodplains means there are now nearly 5 million people and over 2.3 million properties in England and Wales at risk from river or coastal flooding. Climate change is increasing the risk of flooding with rising sea levels and more intense storms. Between July and October 2011, exceptional monsoon rainfall in Thailand caused catastrophic flooding **covering 33% of the country** along the River Chao Phtaya, affecting 8 million people and destroying the employment of 50,000 industrial workers.

Examiner tip

You need up-to-date case studies to support your answers in the exam. Use news websites such as the BBC to research floods as they happen. Make notes on the causes and consequences of the flood event and consider how the flood risk could be mitigated.

Table 5 Types and causes of flooding

Types of flooding	Cause of flooding	Examples
Fluvial flooding is river flooding, when discharge exceeds the capacity of the river channel.	Fluvial flooding occurs either following a long period of steady rainfall over a large area, or because of rapidly melting snow or ice, or on account of a localised intense rainfall event such as a summer thunderstorm.	In July 2007, over 150 mm of rain fell over Wales, western England and parts of Scotland — two to three times the average rainfall. Tewkesbury in the lower Severn valley had 80–90 mm of rain on 20 July, equivalent to almost 2 months' rain in 1 day. This caused extensive flooding. In November 2009, there were floods in Cumbria — over 300 mm of rain fell in 24 hours causing severe flooding in Cockermouth as the rivers Derwent and Cocker overflowed.
Flash or surface-water flooding caused by an intense storm event (sometimes called pluvial flooding as it is the result of excessive rainfall).	Excessive and more localised rainfall that runs off the surface and accumulates in low-lying areas when the soil becomes so saturated that it cannot absorb any more water. Impermeable surfaces in urban areas can contribute to this.	In Boscastle, Cornwall in August 2004, flash floods followed an intense storm in which 185 mm of rain fell over the high ground nearby (24 mm in 15 minutes). River levels rose by 2 m in 1 hour, the banks overflowed and cars were swept out to sea. A 3 m wave of water, held back by debris caught under a bridge, was released suddenly and surged down the main road. Steep valley sides and saturated ground ensured a high amount of surface runoff.
Groundwater flooding	If rainfall continues for a long period, the water table rises to the surface causing overland flow. This type of flood can last for days or weeks.	In 2001 houses in the Somme basin in France were flooded by over 2 m for 2 months. In Lewes, East Sussex, in October 2000 over 600 houses and 200 businesses were flooded. A heavy storm after 3 days of rain caused localised flooding as the ground was waterlogged and drains in the town overflowed. Building on the floodplain and inappropriate agricultural practices had increased the flood risk in the area.
Dam failure flooding	The release of huge amounts of water travelling rapidly is caused by either design failure, or ageing of the construction materials, or a natural process such as an earthquake or landslide.	In June 2007, 700 people were evacuated and part of the M1 motorway was closed when three villages were threatened with the failure of the Ulley dam in south Yorkshire. In 1925, 17 people died when the newly built Coedty dam in Snowdonia broke after 660 mm of rain fell in 5 days. The foundations of the dam had not been adequately constructed.

The physical and human characteristics of a drainage basin that cause flooding

Rivers benefit people by providing natural transport routes and supplying water for domestic, industrial and agricultural use. Almost all of the world's great cities are located on or at the mouths of large rivers. However, rivers in flood can bring damage and destruction and it is necessary to understand how drainage basin systems operate in order to manage them effectively.

Drainage basins with a typically flashy flood hydrograph are those that are more likely to flood. Impermeable land surfaces prevent infiltration so water can only flow overland. Saturated or impermeable soils and geology also limit infiltration with the same effect. The resultant surface runoff is normally contained within river channels unless the input from intense rainfall exceeds the channel capacity.

Changes in land use such as deforestation, poorly planned arable farming or urbanisation also increase the chance of flooding. Any activity that reduces interception removes an important store in the hydrological cycle (i.e. leaf surfaces) which would otherwise slow the progress of water through the system.

Some farming practices increase the rate of runoff:
- ploughing up and down a slope instead of parallel to the contours
- planting winter crops leaving fields bare through the winter
- clearing woodland and hedgerows
- allowing drains and ditches to become blocked with sediment and vegetation

However, urbanisation and industrialisation have the greatest impact on drainage basin hydrology as the land surface is almost completely altered and river systems are managed. Water is a resource for homes, commerce and industry and is stored in reservoirs, treated, used and recycled to meet these needs. At the same time, rivers are managed to enable transport and prevent flooding (Figure 33).

Knowledge check 23

What is a flashy hydrograph?

Examiner tip

Research and learn the details of the Bournemouth flash flood of 18 August 2011 — its causes and consequences. It is a very good example of flash flooding in an urban area that may be useful to know about in the examination.

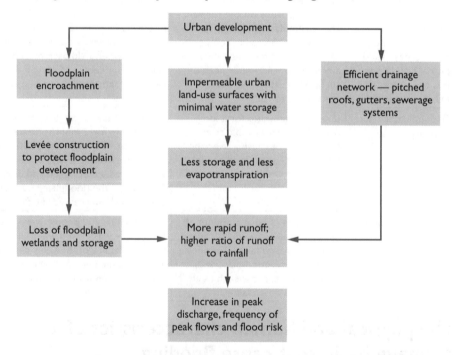

Figure 33 The impact of urban development on runoff

The relative importance of these factors in different flood events

Case study: flooding in England, 2007

May to July 2007 were the wettest months recorded since records began in 1766. More than twice normal rainfall fell across England and Wales, leading to severe flooding in several parts of the country. There were two major flood events: the first in Yorkshire in late June and the second in Gloucestershire in late July, although many other places were affected less severely. The flooding followed particularly high rainfall on 24–25 June and 19–20 July as a result of an unusual weather pattern. This was caused by the position of the **polar front jet stream**, which influences where the weather systems that bring rain to the UK will develop and move. For most of summer 2007 it was travelling south of the UK instead of to the north as is more usual (Figure 34). There were also abnormally high North Atlantic sea-surface temperatures, which increased evaporation and cloud formation.

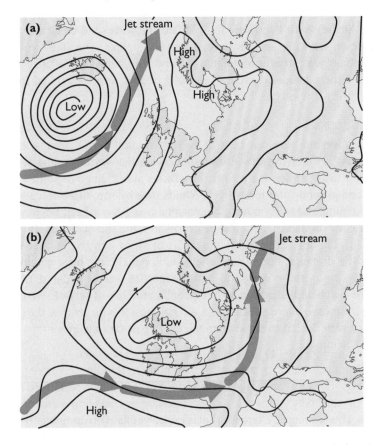

Figure 34 Relative positions of the polar front jet stream: (a) July 2006, (b) July 2007

The volume of rainfall is the main factor in determining whether flooding will occur. Rivers flood when the soil, ground and river channels have insufficient capacity to store excess water. In a typical summer, river, soil moisture and groundwater levels

are usually low, so the ground and soil stores are able to absorb heavy rainfall. The total runoff of rivers in England and Wales for June and July 2007 was well over three times the long-term average.

Surface-water flooding occurs when the rainfall intensity is greater than the rate of infiltration or where surfaces are impermeable. In urban areas, sudden and intense rainfall cannot drain away as quickly as it can in rural areas where the soil is exposed. The speed at which the water is transferred from the surface into water channels is determined by the efficiency of the artificial drainage system. In May and early June 2007, the weather was particularly wet, so river, groundwater and soil moisture levels were already high when the intense rain fell in late June and July. The capacity of urban drainage systems was inadequate to carry the excessive volume of precipitation, which caused flooding.

The June flooding event

Heavy rainfall from severe thunderstorms in northern England in mid-June saturated the ground, so there was little natural storage capacity 10 days later when more heavy rainfall occurred. Four times the average June rainfall fell in places on the North York Moors and in the south Pennines. In Sheffield the Neepsend electricity substation was shut down with the loss of power to 40,000 people. The Ulley reservoir dam near Rotherham was structurally damaged and threatened to collapse because of the weight of the additional water; over 1,000 people were evacuated from nearby villages and the M1 was closed for 40 hours.

Hull: flash flooding

Hull in East Yorkshire received 487% of average rainfall for the area — 256.3 mm in total. On Sunday 24 June the Met Office issued a weather warning and a 24-hour flood incident room was set up for the Yorkshire Dales and Ridings regions. Local rivers were high and groundwater levels in the Wolds rose by 10 m in 24 hours. On 25 June 2007 a deepening and slow-moving depression crossed northern England, bringing sustained heavy rainfall to Lincolnshire, Yorkshire and the Midlands. In Hull 96 mm of rain fell in 24 hours, with rainfall intensities of over 6 mm an hour between 8 a.m. and 5 p.m. The drainage system was unable to cope with the huge volume of water. There was severe flash flooding and 7,800 properties were inundated. Some 10,000 homes were evacuated, affecting 35,000 people, and 95% of schools were damaged. Over 1,300 businesses were affected and golf courses, leisure centres, the race course and theatres were forced to close temporarily. Repair costs were estimated to be in excess of £200 million.

Examiner tip
Find maps of the areas affected by the 2007 floods and compare them with the Environment Agency flood risk areas. Consider the effectiveness of flood risk prediction. Being able to assess information like this may be crucial in examinations.

Hull's low-lying position increases its vulnerability to flooding. Over 90% of its area lies below high-tide level and large parts of the city are built on reclaimed marshland. This makes it especially vulnerable to flash flooding as it has limited natural methods of drainage. The drainage system for Hull is entirely reliant on pumping to remove the water. At first it was thought that blocked gullies (the drains that lead to the sewers) had contributed to the surface-water flooding, but this was subsequently ruled out. It was concluded that the city's drainage system simply could not accommodate the volume of precipitation, causing it to back up and overflow. None of the rivers or streams in Hull overtopped except Setting Dyke to the west of the city — the River Hull did not overtop its banks at any point.

The July flooding event

The Malverns and the Cotswolds received nearly four times average rainfall for July. By late July the soil storage capacity was comparable to that normally experienced in late winter as the ground was still saturated from the previous month's rain. Exceptionally heavy rainfall on 19–20 July caused a major flooding event when a deep depression centred over southeast England and moved slowly northwards. The heaviest rain occurred in Warwickshire, Worcestershire, Gloucestershire, Herefordshire, Shropshire and Oxfordshire. On 20 July, one-and-a-half times the average July rainfall fell around Gloucester in just 1 day, causing widespread flooding. The River Severn is joined by the River Teme just south of Worcester and then by the River Avon at Tewkesbury. The discharges for these three rivers in June to July 2007 were 434% (Teme), 728% (Severn) and 590% (Avon) of long-term average flow. There was massive floodplain inundation and many businesses, homes and vital service sites such as power plants and water works were flooded.

Tewkesbury: fluvial flooding

In July 2007 much of England and Wales received about twice average rainfall for the month. Over 240 mm of rainfall fell in Tewkesbury in Gloucestershire — 531% of average rainfall for the month. Severe storms in mid-July culminated in heavy rainfall on Friday 20 July in the lower Severn catchment. The heavy and prolonged rainfall was caused by a slow-moving area of low pressure and associated frontal system. Tewkesbury is located at the confluence of the rivers Severn and Avon and is surrounded by large areas of floodplain. Consequently, it is frequently flooded. On 20 July both rivers were overwhelmed by the volume of rain that fell in the surrounding areas. All four access roads to the town became impassable. The only remaining access route was by foot along the embankment that used to carry the railway.

In Tewkesbury up to 90 mm of rain was recorded, equivalent to almost 2 months' rain in just 1 day. There was extensive fluvial flooding: 10,000 people were stranded on the M5 and 500 people spent the night at Gloucester railway station as the railway network failed. Properties were flooded either from surface water that could not drain away quickly enough or from overflowing rivers — some were flooded twice. The centre of Tewkesbury town was completely cut off for several days and its water treatment works was inundated, leaving 350,000 people without water for over 2 weeks. Temporary defences were erected at Walham electricity substation, north of Gloucester, to hold back the water and prevent 500,000 people from losing power. However, the nearby Castle Meads electricity substation was shut down as a precaution, leaving 42,000 people without power in Gloucester for up to 24 hours.

Examiner tip

Always try to have an alternative example of a flood. The examiner will credit good use of recent examples.

The physical impacts of flooding within a drainage basin

Flooded river channels have the potential to alter their courses. The increase in the energy of a river, as discharge and velocity increase, enables it to cut through meanders and damage or destroy levées and embankments, both natural and manmade. Morphological change in river channels (i.e. change in the shape of the river) is more likely to happen in uplands nearer to the source of the river. Lowland river channels have lower gradients and more cohesive banks and so maintain their

shape more easily. However, unless river management work has constrained natural processes, the form of meanders changes naturally over time in the middle course of a river and the former paths of rivers can leave detectable scars on the landscape.

When floodplains are inundated the soil becomes waterlogged and crops may be damaged. Macro-invertebrates living in the soil, such as worms and beetles, may die or migrate, which reduces the prey of other species such as birds. Although soils may be eroded in flood conditions, sediment is deposited on a river floodplain as flood water recedes so natural levées may be created alongside the river channel. Floods can also cause contamination of land and soil as low-level household wastes, including sewage, and a wide range of industrial effluents may be dispersed. This can contaminate water supplies and wildlife habitats.

The demographic, economic and social impacts of flooding

The physical consequences of flooding have human impacts including death directly through drowning or indirectly by the spread of disease or the stress of the disruption. The extent to which these demographic consequences impact on a community depends on where people live — the poor are much more vulnerable. Generally, the social costs are usually higher in poorer communities and the economic costs greater in more affluent communities.

Economic costs include loss of or damage to homes and property, loss of crops and farm animals, and damage to or destruction of manufacturing and commercial stock and equipment. During and immediately after floods, communities and businesses are disrupted, causing further economic losses. There are also insurance costs for these losses as well as for rebuilding and repair.

The social disruption caused by floods decreases the quality of life of people and communities. As well as the physical and health dangers of flooding, the psychological impact of the emergency and its aftermath causes longer-term effects. Following the floods in Hull and Tewkesbury in 2007 (see pages 58–59), hundreds of people were unable to return to their homes for many months afterwards — having to live in rented accommodation, with friends and relatives, or in caravans outside their homes. Cleaning up and repairing property, negotiating with builders and insurers, and replacing goods are time-consuming and challenging tasks. There may also be indirect impacts such as a decrease in house values and the fear of further flooding in the future.

Knowledge check 24

What is groundwater flooding?

2.6 How is the flood hazard perceived and managed within a drainage basin?

Different perceptions and awareness of the flood hazard by groups within a drainage basin

Floods cannot always be prevented but they can be managed. The likelihood of flooding is described as the chance or probability of it happening in 1 year. Therefore, there may be a '1 in 100 chance' of flooding in any year in a given location. The following year there will be the same 1 in 100 chance of flooding. Floods may occur in two consecutive years and then not happen again for 200 years, but the level of risk remains the same.

One in eight people in the UK lives on floodplains or in areas at risk from coastal flooding. Perception of this risk varies. Flood risk is not just the likelihood of flooding; it is also the potential impact or consequences of a flood. Flood risk assessments collate information about the hazard and types of risk a flood presents and the likely social, environmental and economic impacts. They are used to assess the need for flood management strategies. The effectiveness of flood control schemes may be tested using cost–benefit analysis.

Flood management organisations assess the potential risks from flooding and then implement appropriate management measures. The Environment Agency has overall responsibility for flood management in England and Wales, in Scotland it is the responsibility of the local authorities and in Northern Ireland the Rivers Agency has this role.

Examiner tip

Look up your own postcode area on the Environment Agency's flood risk map. If you live in or near a flood risk area, how is this risk being managed? Local knowledge may prove valuable in the examination.

Strategies to manage flood hazards

In England and Wales the Environment Agency develops **Catchment Flood Management Plans** (CFMPs) to cover the whole of a river's drainage basin. These plans identify how the impact of floods can be reduced by studying how flood water moves through the drainage basin and developing appropriate management systems to protect property, minimise damage to the landscape and maximise value for money. This integrated approach to flood management goes further than the traditional engineering approach. A wide range of factors is considered including river maintenance, floodplain development, rural land management and the impact of climate change.

Table 6 Flood management measures

Strategy	Measures to reduce flood risk
Reduce the physical hazard using hard engineering	• Flood walls and embankments • River channelisation • Diversion channels and retention basins • Dams and reservoirs
Reduce the physical hazard using soft engineering	• Catchment management • Afforestation • Land-use management • Urban development control • Provision of wetlands
Reduce exposure to the hazard	• Land-use planning • Flood-proofing of buildings
Reduce vulnerability to the hazard	• Flood warning systems • Preparedness and emergency response • Insurance

Examiner tip

Have you got examples of the measures listed in Table 6? Answers without examples do not gain full marks. The River Arun is an example of the use of soft engineering and the provision of wetlands.

Both the risk of flooding and the flood hazard itself require management. Ideally, river management and effective drainage systems can prevent flooding, and defence systems such as embankments and washlands can contain flooding. Sometimes

the volume of runoff overwhelms these systems and the impact of the hazard must be managed. Managed flooding means using natural processes to save areas further downstream. For example, the Amberley Wild Brooks wetland site on the River Arun in Sussex is used to hold back winter floodwater so that settlements and communications downstream are not flooded. Table 6 identifies some flood management measures.

Hard engineering

Hard engineering flood defences, built from hard materials such as concrete or metal, are designed to contain excess river flow. They are expensive and have significant environmental consequences. The ecology of the river and its banks is disrupted and changes to the river channel upstream alter its behaviour downstream.

Examiner tip

Make sure that you have examples of these hard engineering schemes. You may find it useful to research and annotate photographs of the different strategies to help you to remember them.

These flood management measures include:
- modification of the river banks and bed to increase the volume of the river channel
- removal of sediment by dredging to deepen the channel
- removal of boulders to reduce friction and increase river velocity
- construction of dams and weirs to regulate the flow of water
- construction of diversion channels and retention basins to remove flood water
- raising the height of river banks and floodplains to limit the flood

Soft engineering

Soft engineering or 'managed realignment' is designed to work with natural processes. It requires more land but less investment and maintenance. Soft engineering involves managing stages in the hydrological cycle such as increasing interception rates by afforestation or changing agricultural practices. Soft defences can benefit wildlife by providing areas of natural habitat in the rural parts of the floodplains that can withstand occasional flooding, although agriculture may be affected. Embankments may be removed or set back from the river to allow this. In urban areas land uses such as sports fields and parks may be allowed to flood, whereas commercial, industrial or domestic development may be restricted in flood-prone areas.

Flood responses

Flood responses include preparation for a flood event, coping with the immediate impact and dealing with the longer-term consequences.

The Met Office advises on approaching weather systems that are likely to put pressure on drainage networks. It issues three levels of severe weather warning: be aware, be prepared and take action.

The Environment Agency has identified areas generally at risk of flooding because of their proximity to rivers. It issues local flood warnings at three levels:
- Flood Watch — flooding of low-lying land and roads is expected
- Flood Warning — flooding of homes and businesses is expected
- Severe Flood Warning — severe flooding is expected, with extreme danger to life and property

Knowledge check 25

Distinguish between hard and soft engineering.

Advice is given on the organisation's website about what people should do to protect themselves, their families and pets, and their property.

The police coordinate the emergency services during the flood event and people are evacuated from their homes if necessary by the fire and rescue service. The fire service also pumps out flood water. The local authority gives advice and organises accommodation for people who have to leave their homes. There is also a National Flood Forum that offers help and advice on flood protection products and insurance.

The effectiveness of flood management schemes

If heavy or persistent rainfall increases river discharge to bank-full levels, flood management schemes are put to the test. When properties are flooded and people's lives are at risk, flood management may be inadequate. As climate change may increase the risk of localised flooding, flood defences will be put to the test and will sometimes fail.

No matter how careful people are in managing the potential flood risk, there are occasions when flooding is greater than ever before experienced. This photo, taken in Cumbria, shows a warning of flood-damaged bridges following the November 2009 floods. The notice is dated 5 March 2010 because it took 4 months to survey all the damaged bridges and to put up notices.

Summary

- The hydrological processes within the drainage basin and the characteristics of river regimes form the foundation of hydrological change.
- The causes of flooding are both physical and human and their relative importance varies from river to river.
- The impacts of flooding are on the natural environment and on the economic and social life of an area. There are demographic impacts. Impacts vary with the level of development of the affected areas.
- Different groups perceive the flood hazard according to their awareness of flooding.
- There are soft and hard strategies for managing flooding, which do depend on the ability of society to pay for them.
- All flood management schemes must be assessed to see if they are effective.
- You must try to include field studies in your learning programme otherwise question 3(c) becomes difficult to answer.

Questions & Answers

Preparing for the unit test

Geographical terms

It is a good idea to collect together your own list of key terms (your own geographical dictionary) as you come across them in your studies. Also, build up your own bank of simple, effective maps and diagrams. Practise drawing your maps and diagrams because, ideally, you need to be able to spend no more than two minutes drawing each one in the examination. Finally, you will need your own examples to illustrate the points that you make. You will come across examples and diagrams in this book; remember that you can impress by using different, relevant and up-to-date examples rather than the standard ones in your textbooks.

Where to find good examples

Always try to find examples from your own studies in addition to the ones provided by teachers. *Geography Review* and *Geographical* can be good sources of original examples. Websites such as NGfL (**www.ngfl-cymru.org.uk/eng/vtc-home/vtc-aas-home/vtc-as_a-geography**) are useful. Quality newspapers, e.g. *Times, Guardian, Independent* and *Daily Telegraph*, are other sources that provide good examples. But you need to be aware of bias. There is good geography constantly in the media so make use of it. Use news websites and broadcast media to gather information for new case studies, such as floods and earthquakes, as they happen.

The unit test

Timing

There are three questions each made up of three parts. The topics appear in the same order as in the specification. The examination is 1 hour and 30 minutes, and you must answer all three questions. There is, therefore, a mark to be gained every minute. The table below gives you some idea of the timings that you should try to follow. The timings for questions 1 and 2 are identical.

	Activity	Time in minutes
	Read all three questions. Decide on the order of questions to be answered. (The question order below assumes that you answer questions in order.)	1
Q1 and Q2	Study the first resource carefully and answer the 5-mark question Part (a).	5
	Read Part (b) and plan your answer.	2
	Write your answer (probably about one side of prose plus any maps or diagrams).	10
	Read Part (c) and plan your answer.	2
	Write your answer (probably about one side of prose plus any maps or diagrams).	10

	Activity	Time in minutes
Q3	Study the resource(s) for Question 3.	2
	Plan response to part (a).	1
	Write your answer to (a).	6
	Read and plan response to part (b).	1
	Write answer to (b).	8
	Read and plan response to part (c).	1
	Write one-side answer to (c).	9
	Check work.	3
	Total time	**90**

How answers are marked

Markers always assess the *overall* quality of an answer against the mark scheme for that paper. The holistic mark might draw upon qualities from several levels and award a mark that best fits the combination of qualities in the answer. In addition all marks are based on assessment objectives (AOs) which enable marks to be awarded for knowledge, application of that knowledge, and skills. These skills may be the quality of language, or the skills of drawing or interpreting maps, photos, diagrams and text passages.

This levels and marks scheme is used for parts (b) and (c) of questions 1 and 2 and part (c) of question 3.

Level	Marks	Descriptor of the level
3	8–10	Very good knowledge and understanding with a good use of examples. Full descriptions and explanations. There is both breadth and depth to an answer that may draw together a variety of points. Follows the commands in the question. Has a good mini-essay style, well written with a good use of the language of geography.
2	4–7	Some knowledge but not always clear that information/principles are fully understood. Information and explanations are partial. Sound descriptions. There should be more detail. Essay style present but probably lacks paragraphing and introduction/conclusion. Some use of the language of geography.
1	0–3	Basic knowledge with limited understanding. Few examples and if they are present it will be e.g. Africa (often quoted as a country when it is a continent). The material will be simple and probably a very brief one-paragraph response with no development. Slips of grammar and spelling evident.

Quality of written communication

There are no specifically allocated marks for the quality of your writing. However, you should try to use correct punctuation and grammar, structure answers into a logical sequence with a brief introduction and conclusion, and use appropriate geographical terminology.

Managing questions in a minute

All questions have two major components. There are **command words** such as outline, discuss, evaluate, assess. Second, there is the **subject matter**. Use highlighter pens in different colours to emphasise the commands and the topic.

- **Assess** and **evaluate** — weigh up the relative importance of factors. The question may mention one but you should be aware of others. It will generally be used in question 3(c).
- **Compare** — identify similarities and differences or contrasts. When answering questions using compare, it is good mark-gaining practice to use comparative words such as however, larger than, less densely, smaller, steeper and whereas.
- **Contrast** — discuss only the differences.
- **Define** is generally not used in questions. It means 'to give the precise meaning of a term'. However, it does help to be able to define a term used in a question such as 'enhanced greenhouse effect'.
- **Describe** — show what you know is occurring or what a map or diagram shows. It expects you to know what a process does, how it happens, where it occurs and when it occurs. Who or what causes a process and whom it may affect can be part of descriptions.
- **Discuss** — describe and explain the relevant points while building up a strong argument supported by examples.
- **Explain** — say why and how something occurs. It might involve some description but do not rely on description as a means of explanation.
- **Identify** — provide listed points that are supported by examples.
- **Justify** — give reasons.
- **Outline** — state the main points or factors. It is expecting more than two points, with supporting examples.
- **Suggest reasons why** — put forward a plausible explanation from your own knowledge and your broader understanding of the topic.
- **With the aid of a diagram** — draw a diagram, which must be labelled. The axes on a graph must also be labelled.

Examiner comments

In this Questions and Answers section each student's answer is followed by examiner's comments, indicated by the Ⓔ icon. The comments show how the marks have been awarded and highlight specific problems and weaknesses and areas for improvement.

Question 1 Theme 1

(a) Use Figure 1 to describe the influence of climate change on the changing distribution of the mountain ringlet butterfly. (5 marks)

ⓔ This question tests your ability to scan text and gain relevant information from it.

Extinction fear for mountain ringlet butterfly

The only mountain-dwelling species of butterfly in the UK, the mountain ringlet, could be near to extinction in Scotland because of climate change. Experts have warned that warmer temperatures are pushing the butterfly higher up hillsides in search of cooler conditions.

The mountain ringlet faces a very real threat as it retreats further up the hillsides because the warmer weather has made lower habitats unsuitable. The fear is that as climate change continues, this rare species will find no suitable habitats and become extinct in Scotland.

A further consequence of global warming has seen other butterfly species that have died out further south in England surviving in Scotland.

There are already areas in Scotland that are the habitat for butterflies which have become extinct in England and the largely unspoilt landscapes found in the uplands are increasingly important habitats.

Changes in grazing pressure on mountain grasslands and the planting of coniferous woodland have affected the structure and species composition of the mat grasslands — the main habitat of the mountain ringlet — and this will affect the population size of the butterfly.

Butterfly Conservation Scotland (BCS) has asked the public to report sightings of the mountain ringlet.

Figure 1 Changing distribution of the mountain ringlet

(b) Describe and explain *one* example of short-term climate change. (10 marks)

ⓔ The question expects you to describe the change and offer explanation. Defining what is meant by short-term would be a good start.

(c) Examine how the impacts of climate change differ between regions. (10 marks)

ⓔ Ideally you need at least two contrasting impacts but you can examine more. 'Examine' does imply some explanation in this case.

Student A

(a) Figure 1 says that the influence of climate change is causing the butterfly to move higher up the hills in search of cooler conditions **a**. This is due to warmer conditions. The warmer conditions are making lower habitats unsuitable **a**. Figure 1 also shows that the fear is that the butterfly will become extinct because of climate change and increasing temperatures **c**. The species has already died out in the south **b**. Butterflies have become non-existent in England because the temperatures have increased **c**. Figure 1 also shows that changes in grazing pressure on mountain grassland together with tree planting will have an effect on the structure and species composition of mat grasslands **d**. It is where the ringlet lives and it will have an effect on its population size.

ⓔ **5/5 marks awarded.** This is an A-grade answer because it notes **a** vertical movement and latitudinal movement of the species distribution and **b** uses evidence from Figure 1. **c** It also notes changes in temperatures as a major factor. **d** The bit about changing ecosystems is not climate change but human-induced change. The answer uses Figure 1 well and will gain full marks for this answer.

Student B

(a) Figure 1 clearly states that the butterfly is retreating higher due to the influence of climate change raising temperatures. Butterfly numbers are decreasing and the butterflies are migrating to cooler climates **a**. They do not like the planting of woodland which means that they cannot live on grassland . The distribution is changing.

ⓔ **3/5 marks awarded.** This is an answer that suggests that the student knows more than they state. **a** What is given is correct, but no point is developed with real evidence. It is characteristic of a grade-C answer.

ⓔ **On the whole, part (a) of questions 1 and 2 offer an easy chance to gain 5 marks if you are well practised in obtaining information from diagrams, photographs and text.**

Student A

(b) One example of short-term climate change is a period of cooling known as the Little Ice Age **a**. This occurred between 1680 and 1800. The Little Ice Age caused the population of Iceland to drop by half, it wiped out the Viking population of Greenland and in 1780 New York harbour was frozen over **b**.

This period of cooling is believed to be a result of sunspot activity. Sunspot is a large magnetic storm on the surface of the sun. This causes a variation in solar output reaching the Earth. Sunspots emit different amounts of radiation (less) but still emit a large amount of UV rays which cause a reaction in the upper atmosphere resulting in a change in composition. This change in composition and a reduction in solar radiation reaching the Earth causes the planet's average near-surface and atmospheric temperatures to decrease. This is one reason for the period of cooling **c**.

Another reason for the Little Ice Age is the amount of volcanic activity that was happening at the time. Volcanoes emit a large amount of ash and dust into the atmosphere. This causes cooling by blocking out solar radiation. However, the dust only stays in the atmosphere for around 6 months. The eruption also emits large quantities of sulphur in the form of sulphur dioxide. This reacts with water vapour and creates a layer in the atmosphere that reflects solar radiation. This causes a cooling effect on the planet and lasts for over 2 years. This causes climate change for a short period of time **d**.

Another possible reason was that the ocean conveyor shut down. This causes cooling and short-term climate change. During the medieval warm period, large amounts of fresh water entered the oceans. This is believed to have caused the Little Ice Age. The large amounts of fresh water shut down ocean currents and so areas such as Europe stop receiving a warming effect from the Gulf of Mexico, causing cooling in the ocean and leading to short-term climate change **e**.

e **7/10 marks awarded. a** The Little Ice Age is short term in a geological context and was accepted as an answer. Perhaps the answer should have said so. **b** The first paragraph is contextual and vaguely descriptive of effects rather than the whole Little Ice Age. **c** Sunspot activity is a valid explanation. **d** The effects of volcanic activity are shorter term than the Little Ice Age and not related to it very well. **e** The conveyor is a third explanation, although it would have been better to have used the El Niño Southern Oscillation for this answer. It might have been prudent to concentrate more on just the Little Ice Age, although the answer tries to justify the slightly wider explanation. However, there is some good knowledge, which will gain 7 marks.

Student B

(b) An example of short-term climatic change is El Niño and La Niña **a**. These occur every 2 to 7 years and last 1 to 2 years. In a normal year there is low pressure over Australia because of the air rising here and low pressure over Peru/South America due to the air becoming more dense as it descends from the east on account of the trade winds. There is also an upwelling of cold water on the Pacific coastline of Peru **b**.

In an El Niño year trade winds weaken and change to move from the west to the east, causing air to rise over Peru causing an area of low pressure and therefore increasing rainfall **c**. In an El Niño year Peru experiences flooding due to the greater-than-average rainfall from the low pressure. Australia, however, experiences drought in Queensland **d** due to high pressure as air is descending **c**. The upwelling of cold water off Peru is suppressed **c**, which affects the number of phytoplankton and this reduces the number of fish. There are also fewer hurricanes in the southwest USA and the Caribbean. The water is warmer than usual **d**.

However, in a La Niña year the opposite occurs. There is a strong trade wind from the east to the west **b**, causing air to rise in Australia causing flooding in Queensland due to the unusually high amounts of precipitation **c**. The air becomes denser because of descending air over Peru and there is no rain occurring, causing drought in that area **d**. There is an increase in hurricane frequency in the USA and the Caribbean and large amounts of drought in southeastern Australia and South Africa. There is a colder-than-usual feel to the water there **d**.

ⓔ **10/10 marks awarded** **a** The answer begins by letting the reader know what is to be discussed. **b** The description is reasonably comprehensive. **c** Description merges into explanation. **d** The effects are described well. This is a grade-A response for the time available and would gain full marks.

Student A

(c) Climate change affects different regions completely differently. The Inuit colonies **a** are struggling due to large ice melts **d**, less animals to hunt and poor snow to build with. This means that some are having to migrate and there is a loss of culture, poor standard of life as some find it hard to build their igloo's **d**. Ice and snowmelt also means that some pathways are not able to be used in some settlements and will have to travel a long way round.

In the Kiribati Islands in the middle of the Pacific on the equator **a** there are huge problems as they are all on average 3 m above sea level. With rising sea levels occurring fishing settlements that are built right up to the waters edge will suffer as in 2005 tides were recorded at 2.85 m high, which is very serious. The first row of plants and coconut trees have been eroded as well as the intrusion of salt water affecting the water system and crops. As food sources are depressing and there are many remote islands there is panic and people are starting to out-migrate to New Zealand and Australia, which cannot deal with an influx of thousands more people **e f**. If worst comes **d** and the islands are engulfed over 85,000 people would have to be relocated with the cultural loss of the Kiribati Islands as people leave due to the decreasing standard of living. This would cost the population lots of **d** their savings and some would not be able to move **b**.

In Kenya **a** there have been droughts for long periods. One lasted 3 years with farmers and nomads dying with all their cattle as wells dried up and they had little money and little education to set up a new life elsewhere **b c f**.

ⓔ **6/10 marks awarded. a** Three cases of change are outlined but **b** the student does not really tell us how the changes differ. **c** The answer describes changes and does not link the three case studies together. **d** The quality of language does not help this student — an apostrophe is in the wrong place and there are some low-level words such as 'lots of'. This response could do with named cases of places in north Canada and island names in the Pacific. **e** It is exaggerated when discussing the impact on Australasia. **f** Ideas are jumbled together and not necessarily very logical. There are elements that are correct but they are not linked strongly to climate change. The answer lacks comparative examination and the student thinks that by just describing three sets of impacts he/she has examined the differences. The answer is characteristic of a grade C candidate and would gain 6 marks.

ⓔ **Student A scores 18/25 overall, which is a B grade. Full marks on part (a) made up for a weaker performance on parts (b) and (c).**

Student B

(c) One region that is affected by climate change is the land around the world's largest deltas **a**. Here many people live because of the fertile land which provides a livelihood for people in agriculture. As climate change increases and temperatures increase, these regions are increasingly vulnerable to flooding both from the rivers and from the rising sea levels caused by ice melt. The impact here is that agricultural land will be destroyed, meaning loss of both food and income because of the loss of crops **b**. This means that many people will be displaced. In the Ganges Delta of Bangladesh mangrove swamps are already being destroyed **a**. Further increase in sea level would destroy 20% of the country's land and displace 40 million people. In the Nile delta in Egypt a 1 m rise in sea level would affect 7 million people and 15% of the habitable land **a**. Here the impacts are social and demographic **b**. Displacement can cause great distress as people are forced to leave their homes. Once they move elsewhere it may be difficult to find work and so this can lead to poverty **b**. The demographic impacts are that the areas they move to may suffer from overpopulation as the population distribution changes as does the population density **c**.

    Another region that will suffer from climate change is low-lying islands such as the Netherlands **a**. Here the impacts are different because the Netherlands is an MEDC and Bangladesh is an LEDC. There are different economic, social and environmental impacts. Half of the Netherlands is below sea level and densely populated. Although these areas are already defended, a 1 m rise in sea level would cost $12 billion to protect **b**. This means that the impacts are economic. People in MEDCs are more likely to have insurance on their property and because of the high standard of living in these areas they will be less inclined to move away **b**. This means it is important for the government to invest in these sea defences **c**.

🅔 **9/10 marks awarded. a** This is a good response because it contrasts areas where the standard of living differs considerably, discussing three deltas in areas of contrasting economic wealth. **b** The answer recognises different impacts: economic, social and environmental. **c** It uses the language of geography well. It would help if some actual places within the countries were named. Nevertheless, this is a grade-A response, gaining 9 marks.

🅔 **Student B did not start well but has scored 22/25, which is an A performance.**

Question 2 **Theme 2**

(a) Describe the local impacts of earthquake activity shown in Figure 2.　　　　(5 marks)

The cathedral in the centre of the town of L'Aquila

The village of Onna

Figure 2 Impacts of the earthquake in central Italy, 6 April 2009

ⓔ This is testing your ability to find information on photographs.

(b) Discuss some of the social impacts that are the result of tectonic activity.   (10 marks)

ⓔ The question does not expect every social impact, but make sure that you keep to social impacts.

(c) Outline two different strategies used to manage tectonic hazards.   (10 marks)

ⓔ The two strategies must be different. Two strategies relating to one hazard would be acceptable, although the question allows you to take your examples from different tectonic hazards.

Student A

(a) Earthquakes cause shock waves that travel through the Earth causing a variety of crustal movement. Figure 2 shows the major result of this movement: building collapse **a, b**. The cathedral in L'Aquila has suffered great damage, roof collapse and cracking down the sides of the buildings, yet the statue beside it shows minimal damage, probably due to differences in building material **b**. The village of Onna shows even greater damage **a** — one building is almost totally collapsed, burying vehicles and people, and debris is obstructing transport routes **b**. Telephone lines have gone down and electricity cables have been damaged **b**.

ⓔ **5/5 marks awarded. a** This is an example of an excellent use being made of the resource photographs. **b** Five impacts are described and referenced. This will gain full marks.

Student B

(a) Figure 2 shows how an earthquake can have an impact on local surroundings. There is much damage **a**. Cathedrals are big tourist attractions **a**. Earthquakes mean that less tourists will visit it. The second photo shows destroyed houses **a**. Everything is lost, which has other impacts **b**. People are displaced. A car is buried beneath the rubble and this person has lost their transport **a**.

ⓔ **3/5 marks awarded. a** This response uses the photographs in a vague way. **b** It does not always make the link between the evidence and the statements about impacts; nor does it use evidence to make clear points. The response will just gain 3 marks.

Student A

ⓔ Student A's plan has not been printed here. Writing a small essay plan is a good idea.

(b) Tectonic activity creates two major hazards — volcanic eruptions and earthquakes — which have great social impacts on a variety of scales **a**. They affect health, education, employment, water and food supplies, property and livelihood, and have individual impacts on emotional stress **a**. For example, in the aftermath of the 1985 eruption of Nevada del Ruiz in Colombia, the town of Armero was flooded with lahars, which buried the town in 6 m of mud and killed 70% of the population — about 20,000 deaths. It totally destroyed most things in its path including homes, hospitals, schools, industry and farmland — some people lost everything and the economy struggled to pay for the impacts which had long-term effects as well as short-term ones such as the spread of disease and lack of food, shelter and safe water supply **c**. The Mexican earthquake of 1995 also caused severe social impacts as it destroyed Mexico's biggest hospital, Centro Medico, which the country could not rebuild as it suffered $96 billion of foreign debt **c**. Some argued this had worse long-term impacts than the thousands of deaths **b** caused by the earthquake itself.

The Asian tsunami of 2004 left 900 children orphaned in Sri Lanka and over 220,000 deaths across the Pacific area **a b d**. The impacts of the tsunami caused by a 9.0 Richter scale earthquake were felt as far as Somalia and Kenya. Millions were left homeless with no emergency supplies and no means of getting aid or clean water, which encouraged the spread of water-borne diseases such as cholera and also malnutrition and starvation **c e**. Increased levels of depression were also a widespread impact of the devastating tsunami.

ⓔ 7/10 marks awarded. a This answer shows that the student knows that there is a wide variety of impacts. **b** You are expected to know the difference between demographic impacts and social impacts and this student, in discussing deaths, does drift into the demographic, which is not strictly required here. Some credit is given for including demographic within social. **c** However, more credit is gained by the references to health, effects on the mental state of people and the loss of homes. **d** Regarding the Asian tsunami, it should read Indian Ocean rather than Pacific Ocean. **e** Some sentences such as the penultimate one in the last paragraph are conflating too many ideas into one sentence and suggest that the student is trying to keep to time rather than thinking about what they are writing. The response will gain 7 marks, which is just an A.

Student B

(b) Tectonic activity can have different social impacts on different societies. In Japan once a year the people take part in an earthquake drill to teach the young and prepare everyone for the earthquake. Japan is an MEDC and so it can afford this. It means that the country is prepared. As well as people, structures are built earthquake-proof. This means that if there is an earthquake, there are less impacts on society. However, some impacts include transport links being closed; people can die and be injured. Japan is at risk from short-term impacts **b c**.

However, in LEDCs the social impacts are greater. In Haiti in January 2010, 230,000 people were killed and many were made homeless. Buildings were poorly built and collapsed easily. The government building collapsed. The earthquake was managed poorly. Emergency services were slow to react and so more people died **c**. Survivors relied on international aid for food and water, which there was not enough of; people starved and disease spread. Haiti is using aid to rebuild the country **a c**.

People live near volcanoes as they offer many advantages. Agriculture thrives in the volcanic ground so there are many farmers. Volcanoes such as Mount St Helens were tourist attractions. This offered jobs to local people **b**.

ⓔ 5/10 marks awarded. This is an example of someone who is writing all they know, possibly answering a question that was completed for homework. **a** In particular, the second paragraph is descriptive of an event rather than highlighting the social aspects. **b** It would have been better to exclude the paragraph on volcanic activity, which focuses on economic impacts (the marker will just ignore that paragraph). Economic impacts are there in the first paragraph. **c** The impacts need to be less generalised and more specific. This response will obtain a level 2 score of 5 marks (grade C).

Student A

(c) In the Philippines, after Mount Pinatubo **a** started showing signs of activity after 600 years of dormancy, Philippinean and American scientists flocked to the area to monitor the processes within the volcano. Historical evidence showed them that huge eruptions had happened in the past and seismology, laser tracking and gas omissions all indicated an imminent eruption. Hazard maps were drawn up to indicate the areas most at risk, which were subject to evacuation in accordance with the monitoring reports for the volcano. Approximately 70,000 people were displaced in the evacuation, but when the volcano violently erupted in 1991 spreading pyroclastic flows across 40,000 km^2 and ash clouds which reached Vietnam, only 800 people died, mainly due to secondary hazards such as lahars and disease. The mass evacuation of villages and towns on the slopes of Pinatubo saved thousands of lives, as did accurate prediction and good communications between scientists, the military and the government officials who responded quickly and effectively. **b**

During the last eruption of Mount Etna **a** in the 1990s lava flows were attempted to be controlled. In the path of the lava flows were villages further down the slopes that faced total destruction from the advancing lava flow if nothing was done. Engineers constructed a plan that involved using explosive dynamite to redirect the flow into a manmade canal and thus diverting the lava into an uninhabited area and away from its course towards the villages. The solidified bank of the lava channel was to be broken by explosives so the lava could flow into the new channel. This was done with some success as 30% of the lava was diverted but there was still some damage.

e **9/10 marks awarded. a** This mini essay discusses two case studies of strategies to manage volcanic eruptions. It is a pity that the student did not write a short sentence to introduce the topic and a short concluding sentence to sum up the two studies. **b** The two studies could have been linked by a sentence such as 'Another strategy involves reacting to lava flows.' The examples are good and therefore enable the student to gain a level 3 mark. It is level 3 for quality of language and overall gains 9 marks.

e **Student A has scored 21/25, which is a very good A grade, by answering the question and supporting his/her responses. Some examples are well developed and have good detail.**

Student B

(c) In Hawaii there is a tsunami warning centre. When an earthquake occurs in the Pacific plate boundaries it will take a few hours to reach Hawaii **d**. The following tsunami is monitored as it moves towards the islands **a**. The beaches are evacuated and people moved to higher grounds if the tsunami is a threat **b**.

In Japan children are taught from an early age on how to deal with an earthquake **a**. The country has an annual 'earthquake day' which helps people prepare for an earthquake. The government has been moved out of Tokyo as this is an area at great risk of tectonic activity. Buildings are built to deal with an earthquake such as some with giant pendulums inside to counteract the swaying. All families and households have an earthquake emergency kit. All this is done to minimise damage to the economy, loss of life and injuries **b c**.

ⓔ 5/10 marks awarded. **a** There are two strategies as required by the question although **b** both are rather brief and, **c** in the case of Japan, a group of strategies rather than a developed single strategy. The tsunami case is in need of more explanation of how the system works: the recording buoys, the sirens and the procedures. **d** This response illustrates lack of quality of expression especially at the start when it is the tsunami and not the earthquake reaching the islands. Always remember to write clearly. It is a level 2 answer for both content and quality of language and would achieve 5 marks.

ⓔ **Student B scores 13/25, which is a low grade C. The answers lack detailed explanations and support is vague. The quality of written language is below what examiners would expect.**

Tackling question 3(c)

Question 3(c) in your examination will expect you to be able to write about **your own study of a topic in physical geography**. You will be expected to gather information, know sources of information, have acquired the skills of mapping and field observation, interpret information, realising that there may be bias, and draw conclusions from that information. Hydrological studies provide the ideal opportunity for you to develop these skills through a field exercise concerned with hydrological change. Your approach to this part of your study of Changing Physical Environments should follow a modified sequence of enquiry.

ⓔ The best way to improve your grade is to concentrate on improving answers to question 3. Many students risk throwing away a good grade by giving less time to question 3 or by not answering parts of the question. This is a dangerous activity. For example, if you fail to provide an answer to question 3(b), that subtracts 8 marks — which will make at least a grade difference to your overall mark.

Hydrological enquiries that you could attempt

(1) An investigation of people's perception of the flood hazard risk in an area.

(2) An investigation into the impacts of flooding in a recently flooded area.

(3) A study of seasonal variations in discharge along a stretch of a stream.

(4) To what extent does the River X fit the Bradshaw model?

(5) An investigation of how river characteristics change downstream.

(6) How can flooding on the River X around town X be managed?

(7) An investigation into the variables that will affect flooding on the River X.

(8) How do changes in cross-profile vary downstream?

(9) What are the effects of either culverting or channelisation on stream characteristics?

(10) An examination of the Flood Watch system on the River X.

The sequence of enquiry

(a) Plan the enquiry

Planning is a key element in carrying out an enquiry and features in questions in the examination. It is the whole process of deciding on the topic, reading up and organising the study.

(1) Read about the topic in textbooks and in articles in journals and magazines such as *Geography Review*.

(2) Decide on a topic that is either an area of investigation (e.g. attitudes to a planned flood defence scheme and distance from the river) or a question that you wish to answer (e.g. does velocity increase downstream along a 5 km stretch of the river?).

(3) Plan how you will get the information and data from primary sources (e.g. fieldwork and questionnaires) and secondary sources (e.g. flood risk maps) that will enable you to answer your investigation topic. Prepare the relevant data collection sheets.

(4) Carry out a risk assessment in collaboration with your tutor. Always let people know where you are going and for how long. Have you got the right equipment both to gather data and to ensure your own safety? Have you got permission to enter the areas that you wish to study?

(5) Many topics in physical geography require group work. There is no harm in doing a group project as long as you and your colleagues all know what you are doing and why. An ideal group is 2–5 persons. Above that size you may find some people just leave you to do the work for them.

(b) Carry out the fieldwork

(1) Go to the field locations and record any data on your data collection sheets.

(2) Be aware of the faults and errors in your data collection.

(3) Return to collect further information if necessary.

(c) Process the information

(1) Graph and map the data collected. Understand what the information that you have gathered actually says about the topic or question you are studying in the field. How accurate was your data gathering?

(2) Be able to process data statistically. You will *not* be asked about statistical processing in the examination, but you might need to know what the statistical processes show.

(d) Write 1,000 words on your study

(1) Organise the information and data so that it answers your question.

(2) Draw conclusions from your results and write them down as in a sequence of enquiry essay. Examiners have found that people who have not gone through the whole process, including drafting an essay on the investigation and its results, do not score as well as those who follow this advice.

(e) Evaluate your work

(1) Evaluate each stage of your investigation so that you can answer questions about the sequence of enquiry in an examination.

(2) Evaluate your evidence and conclusions in relation to what you read and were taught at the outset and what you have found. This should include altering the topic or question that you asked so that others could gain a better understanding of the subject in the future.

Questions you could be asked on your study

The following questions are examples that address the sequence of enquiry above.

Stage (a)

- Outline how you planned your study into a changing physical environment.
- Outline the planning stages of an investigation that you have undertaken into a changing physical environment.

Stage (b)

- Assess the accuracy of your methods of data collection in an investigation that you have carried out into a changing physical environment.
- What are the methods of data collection that can be used in an investigation into a changing physical environment?

Stage (c)

- Discuss the strengths and weaknesses of the methods of presenting data used in your investigation into a changing physical environment.
- Examine the relative merits of the various methods used to portray information in your study of a changing physical environment.

Stage (d)

- Discuss the main conclusions of an investigation into a changing physical environment that you have undertaken.
- Discuss the reasons why the conclusions to your investigation into a changing physical environment backed up what you read when planning the study.

Stage (e)

- To what extent is it possible to increase the validity of the conclusions of your investigation into a changing physical environment?
- Discuss the value of using a sequence of enquiry when investigating a changing physical environment.

Question 3 **Research including fieldwork**

(a) Describe the changes resulting from the 2004 tsunami on the area shown in Figure 3.　　(7 marks)

Figure 3 Before and after the tsunami of 26 December 2004, Khao Lak, Thailand

🄮 This question expects you to demonstrate your ability to monitor change on air photographs.

(b) Examine the value of photographs and field sketches in the investigation of changing physical environments. (8 marks)

ⓔ This question follows on from the opening question and asks you to demonstrate your knowledge of the value of photos and sketches. It expects you to have examples to illustrate your answer. It is used solely for Student A.

(b) Outline how you could collect information on people's views about the Afon Adda flood prevention scheme. (8 marks)

ⓔ This question, used for Student B, is from a different examination. It is asking you about how to collect data — stage (b) of the sequence of enquiry on page 79.

ⓔ **There are two examples of question (b) here, but you will only have to answer one part (b) question in the examination.**

(c) Discuss the strengths and weaknesses of the methods of presenting data used in your own investigation into a changing physical environment. (10 marks)

ⓔ You should state clearly the question you have investigated. This is about data presentation and the strengths and weaknesses that you found when you wrote up your investigation.

Student A

(a) Figure 3 shows that the tsunami has caused a reduction in the amount of vegetation in the area **a**. The river channel is also wider than before due to the large influx of water **a**. As a result of the tsunami, houses have been destroyed as well as the beach **a**. The 'after' photo **b** shows a large amount of beach erosion because it is no longer there **a**.

The 'after' photo shows a large amount of deposition at the mouth of the river **b**. A small island has also been formed where the river has cut a piece of land away from the mainland **a**. It also shows that the general richness of the area has declined due to the tsunami **c**. The land is now covered in mud and debris. The figure shows some flooding as a result of the 2004 tsunami **b**.

ⓔ **6/7 marks awarded. a** The answer refers to several changes **b** using evidence from the two photographs. It would have scored more if it had said where on the photos the evidence could be seen. The statement about richness needed support. It is nevertheless a grade A response, scoring 6 marks.

Student B

(a) The most notable change seen in Figure 3 is that after the tsunami the wide sandy beach that was the main tourist attraction has been completely submerged **a**.

The majority of the residential areas have been destroyed **a**. Less than half the buildings have remained upright but I do not know what state they are in.

There has evidently been a huge damage **a** to the economy as the hotel complexes, residential areas and the beach have suffered extreme damage which will hit the tourist industry hard **b**.

Environmentally, there has been a huge amount of deforestation **a** caused by the tsunami. It is clear to see that over two-thirds of the vegetation has been uprooted by the wave **b**.

This has left vast areas of soil exposed and without vegetation roots and it will no doubt be subject to extreme soil erosion **b**.

The sea is much murkier in the 2004 photograph. This infers that there is a lot of sediment being carried that will have to be deposited somewhere.

(e) **4/7 marks awarded. a** The theme of much of this response is damage and as a result it deals with changes in passing rather than focusing on changes. **b** It could be more detailed by referring to areas on the photographs rather than expecting the examiner to know which part is being discussed. The short paragraphs detract from the flow of the answer. It is a level 2 response that will obtain 4 marks.

Student A

(b) The value of photographs and field sketches is high in an investigation **a**. Sketches and photographs can show explanations that may be difficult to put into words **b**. Photographs can be compared with other photographs **c**. This can easily show any changes you are looking for. This means the value of photographs and sketches is high because it can help in the main and sub-hypotheses **d**.

Photographs and sketches have a high value because they can help back up any conclusions made and so increase the validity and reliability of conclusions **e**. This increases the value of photographs even further because data collected needs to be accurate and reliable to be able to answer the hypothesis in a valid and correct manner.

Photographs and field sketches are valuable because they can be analysed at a later date, which reduces the chance of missing any data or recordings out. They are also valuable for showing changes, which is the whole aim of the investigation **c**. They are also easy to interpret and understand as they show clearly what you are trying to describe or explain **e**.

(e) **4/8 marks awarded. a** The first sentence is an unrealistic claim. **b** The second sentence lacks evidence and is rather unfounded. **c** The point about changes, no doubt following on from the response to part (a), is valid and would have been better if an example of changes shown in photographs had been given. **d** The last sentence of the first paragraph says very little. **e** The answer needs reference to the actual use of photographs and sketches rather than the hypothetical case made here. The answer makes no real reference to field sketches — the second part of the question — and, for that reason alone, gains half-marks.

Student B

(b) To collect information on people's views about the scheme, first a questionnaire should be devised **b**. This would make it easy to compare everyone's views because the answers to specific questions could be compared. It would break down people's opinions and so make the results easier to work with. Questions could be 'Do you feel the flood prevention scheme is effective?' **d** or 'Do you have any suggestions for improvement?' **d** Using questions guides the person so that they know clearly what to say.

    Another way in which the data should be collected is by a stratified sample **a**. It would be too time consuming to give questions to every inhabitant of the town. A stratified sample means asking a set amount of people from different age groups and different genders so that the results show the views of the whole of the population **a**. Doing a random sample would not be effective; for example, more males from age groups 50+ could be asked to fill out the questionnaire than other age groups and women and so the results would not be fair **a c**.

ⓔ **6/8 marks awarded.** This answer is disjointed because **a** sampling techniques should have been placed before **b** the nature of the questionnaire. **c** The student does not suggest that there should be a pilot of the questionnaire. **d** The two suggested questions are valid but they should be more rigorous. Perhaps a tick box set of reasons could have been given. Because there are valid potential questions and reference to sampling, this answer will gain 6 marks.

Student A

(c) The hypothesis of my investigation was 'Does discharge of the River X change from source to mouth?' **a**

    To present data I used recording sheets and first results tables. This is a good method of presenting data used in an investigation because it shows all the raw data that is collected and so provides depth. However, recording sheets are not good as they can be difficult to interpret and they can make it hard to identify trends **e**.

    I used bar graphs to display discharge **b**. This method is good because bar charts are easy to construct. They are easily understood and they show trends effectively. However, they do make it difficult to identify rates of change. But they are easy to compare.

    Line graphs were also used to display data. This was a good method because they are easy to construct, easy to understand and easy to compare. Another strength is that they show trends clearly. The main strength is that they show rates of change, which is very useful. **c**

    To show depth and width of the river I used cross-sectional diagrams. These are a good method of displaying data because they are easy to construct and compare. However, they are not good because the width and depth they show is not accurate. The river bed was not flat and so drawing its shape accurately is difficult.

    The final results table was a good way of displaying data because it showed the data we wanted to collect. It showed an average at each site I investigated and it was fairly easy to compare with other sites. However, it showed averages and not raw data, which means it estimates. **d**

ⓔ **5/10 marks awarded. a** The subject of the investigation is stated clearly. **b** Bar graphs, line graphs, cross-sections and a results table are mentioned but with little linkage to the actual study. What data were displayed by the line graph? **c** The emphasis is frequently about ease of drawing rather than their use to explain changes in the river from source to mouth. **d** Similar reference is made to ease of interpreting but the reader does not know what was being interpreted as the content of the graphs is not stated. Therefore, the strengths are repetitious. **e** Weaknesses are given little consideration except in the discussion of the tables (we do not know what was in the tables). The essay is level 2 and gains 5 marks.

ⓔ **Student A has achieved 15 marks, which would gain a grade C. The performance was variable and needed to pay greater attention to the questions and provide support from the real world.**

Student B

(c) I investigated 'How the methods of flood management used around the X (river named) tributary in X (county named) are influenced by land use.' **a**

The most effective means of presenting the data I collected was compiling a land-use GIS map. This was very effective as it was visual and clearly portrayed the three different sets of data **b**. The basic map had river location on which proportional flow arrows were used to clearly and visually show the river discharge at six different locations **c**. Superimposed on this were pictures of each land-use type with adjacent bar charts to show a people count. This was effective at showing multiple pieces of data **c**, but as it was created by hand on a fairly small scale there were inaccuracies, particularly when creating proportional flow arrows **c**.

Another means of presenting data were radar graphs. This showed the data personally compiled on my opinion according to the aesthetics, strength, use of materials, etc. of the management techniques **c**. These were again visual which was good as you could see instantly which were particularly successful methods and which were not **d**. However, something like a line graph would have been more effective at showing all the data simultaneously, which would have made it easier for direct comparison **d**.

Tables were also used to present data **b** and, although they contain all the necessary data, they are not visual and they can be hard to draw direct comparisons. However, they enable you to calculate means and averages.

ⓔ **7/10 marks awarded. a** The title is given. **b** The study is a valid one although the exact nature of the data is not clear. **c** Various presentation methods are discussed and evaluated in a general format. **d** There is some evaluation. To get more marks try to avoid repeating the same phrase or same reason for using a technique. This response would gain 7 marks.

ⓔ **Student B's marks are a compilation from two students. The marks total 17/25, which is a B grade.**

Knowledge check answers

1 It is a community of plants and animals within an environment such as the tundra or savanna grasslands. Biomes are large-scale ecosystems.

2 It is the circular movement of a mass of air oscillating back and forth across the Pacific Ocean in the tropics and subtropics. In 1969 Bjerknes noted that trade winds across the tropical Pacific flow from east to west. To complete the loop, air rises above the western Pacific and flows back east at high altitudes, to descend over the eastern Pacific. He called it the Walker circulation and established its link with the oceanic changes of El Niño and La Niña.

3 It is the boundary between a shallow, warm, upper layer and a deep, cold layer of the ocean.

4 Carbon dioxide is increasing but varies globally. Methane is also increasing and has a global warming potential 21 times greater than carbon dioxide. Nitrogen oxides contribute a small percentage of the gases but have a greater global warming potential. CFCs (which have also damaged the ozone layer higher in the atmosphere).

5 The long-term changes are caused by natural influences such as variation in solar radiation and the effects of the Milankovitch cycles. Short-term changes are caused by volcanic eruptions (global dimming) and, in recent decades, by human activity increasing the proportion of greenhouse gases in the atmosphere.

6 The volume of ice has decreased by nearly 10 billion cubic metres in 24 years. The reduction in the volume of glaciers has been fairly steady, apart from a slight increase in the late 1980s.

7 Eustatic change is the change in the volume of water in the sea whereas isostatic change is the movement up or down of the land. Both affect the sea level relative to land.

8 Natural forcing involves changes in global temperature caused by natural processes whereas anthropogenic forcing involves changes that result from human activity, especially in the past two centuries.

9 Positive feedback occurs in an open system when a change in one of the variables sets in motion a snowball effect, whereas negative feedback is when the system adjusts to counter the effect of the initial change.

10 El Niño is the warming of the eastern Pacific Ocean following the transfer of warm water from the western Pacific, which alters the climate and ocean ecosystem characteristics in and off Peru. La Niña is the reverse climate for the eastern Pacific when warm water is pushed westwards resulting in changed climatic conditions in Indonesia and Australia.

11 It is an area that has the capacity to absorb large volumes of carbon dioxide, such as the major global forest areas. Some deep ocean areas are also carbon sinks.

12 Replacing fossil fuels with biomass. Renewable energy can include solar panels on domestic roofs. Energy conservation measures could include cavity wall insulation or more efficient car engines. Transport policies such as in Copenhagen where 60% commute by bicycle. Forestry is also encouraged by various government schemes.

13 Interest groups are groups of like-minded individuals who may become a pressure group when their particular interest is threatened. Birdwatchers in a wetland have an interest and become a pressure group if a road is planned through the area.

14 Introduced in 2010 by the UK government it aims to enable private firms to offer consumers energy efficiency improvements to homes and to community facilities. Payment will be recouped through energy bills.

15 Check your answer by studying Figure 19. You should also learn the type of boundary where two named plates meet — for example, the boundary between the Nazca plate and the South American plate is destructive.

16 Constructive: where plates are moving away from one another, e.g. the mid-Atlantic ridge. Destructive: where plates are moving towards one another, e.g. off western Sumatra. Conservative: where plates are sliding past one another, e.g. the San Andreas fault.

17 It is a very hot mass of lava, ash, water and gases flowing rapidly downhill from an explosive vent in the form of a pyroclastic flow. It is ejected in the most dangerous type of volcanic eruption for human life.

18 A Jokalaup is the melting of snow and ice over a vent, resulting in flash floods. Topography will guide the course of the water.

19 Closed systems have no import or export of materials or energy across the boundary of the system, e.g the hydrological cycle, whereas in an open system imports and exports across the boundary do occur so that the system may continue, e.g. in a drainage basin.

20 Infiltration is the movement of water from the ground into the soil. Percolation is the downward movement of water in the soil and rocks under gravity. Throughflow is the sub-surface movement in a downslope direction, especially above impermeable rocks towards springs.

21 Perennial flow is fed by a constant supply of groundwater. Intermittent flow is when flow varies due to fluctuation in the inputs of water from rain or groundwater, as is the case in many chalk streams that flow in winter or after spells of prolonged and heavy rain, but rarely in the summer months. Ephemeral flow occurs during and after a rainstorm but only for a few hours or days; such streams are typically found in more arid regions.

22 The process by which precipitation is prevented from reaching the ground by vegetation and buildings. The water is stored and falls to the ground by stemflow and drip or is taken into the drains, so altering the pattern of flows.

23 It is a hydrograph that shows a rapid or steeply rising onset of water flow followed by a brief high peak and then a fairly fast decline.

24 This occurs if rainfall continues for a long period, and the water table rises to the surface causing overland flow. The Cockermouth floods in 2009 were caused by prolonged, excessive rainfall in the Lake District.

25 Soft engineering works with the natural processes for the benefit of an area and its people, for example, by permitting controlled flooding. Hard engineering attempts to control the processes by physically altering the stream channel — for example, by widening, channelisation and diversion.

Page numbers in **bold** refer to **key term definitions**